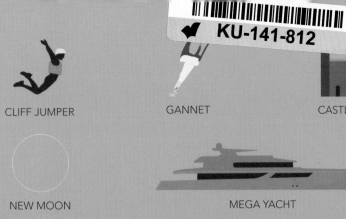

CLIFF JUMPER

GANNET

CASTLE

NEW MOON

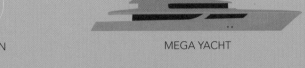

MEGA YACHT

FIRST QUARTER

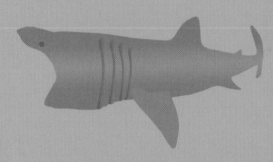

FULL MOON

BASKING SHARK

THIRD QUARTER

STREAM

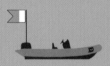

DIVER BELOW

BINOCULAR VISION

DIVE MASK VISION

SCUBA DIVER

ROYAL NAVY WARSHIP

PLAYBOATER

1 FATHOM
6 FEET
1.82 METRES

WIND

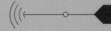

SWELL

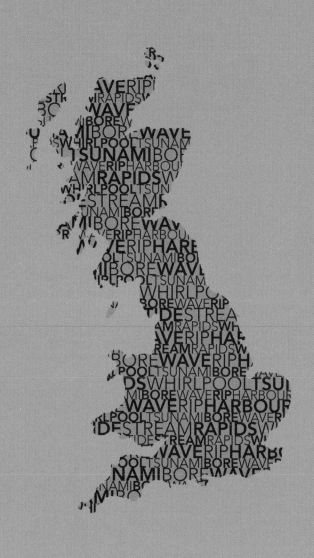

THE
BOOK
OF
TIDES

A JOURNEY THROUGH
THE COASTAL WATERS OF OUR ISLAND

WILLIAM THOMSON

Quercus

First published in Great Britain in 2016 by

Quercus Editions Ltd
Carmelite House
50 Victoria Embankment
London EC4Y 0DZ

An Hachette UK company

A CIP catalogue record for this book is available
from the British Library

HB ISBN 978 1 78648 079 8
EBOOK ISBN 978 1 78648 080 4

10 9 8 7 6 5 4 3

Printed and bound in China

This book can be traced directly back to the day I first appreciated the power of *stream* [Chapter 2]. It was a hot summer's afternoon and I was part of a lifeboat crew skimming across the English Channel at 38 knots. Out there, 19 km from Britain and slightly closer to Europe, a yacht was being swept into the busiest shipping lane in the world. With no wind to power the sails and a rope tangled around the propeller, the *Lionheart* was at the mercy of stream.

Water flows both ways through the English Channel – six hours at a time. During each six-hour period the flow speeds up for three hours then slows down for the next three. The time it is slowest is called 'slack water' [always the same time before and after high tide] and this is when the direction of stream changes. When the submerged rope tangled itself around the yacht's propeller the stream was flowing south towards the shipping lane, where there was a very real danger she would be run over by a mega-tanker. So our job was to get her out of there. As an [extremely inexperienced] sailor my role was to board the yacht, tie all the knots to secure her to the lifeboat and then take the helm. It was an afternoon well spent, especially considering my colour blindness usually restricted me to shore-based jobs with the RNLI.

That day highlighted the perils of stream but they do not just occur far out at sea – the same principle happens just off the beach. A visitor to Deal in Kent learnt this the embarrassing way. It was another hot summer's day and the elderly lady decided to go for a dip in a very skimpy swimsuit. When she got out she discovered her bag had been stolen from the beach. Photographs in the local paper showed her running around the high street trying to find the culprit. What she didn't realise is the seemingly still sea had swept her about a kilometre down the beach and her bag was exactly where she had left it.

It's not just visitors who are surprised by the motions of the sea. Talking to locals, I noticed how few people were aware of the cycle of stream. Many wondered why sometimes they got swept one way and other times the other. Some feared a receding tide would suck them out to sea. I resolved to combine the knowledge I'd acquired with my training as an architectural draughtsman and illustrator to create a simple map format that the locals and their visitors could understand. Often they were sailors and kayakers wanting to know how to use the waters safely, but they were also people wanting to celebrate the unique landscape and seascape around them.

The maps started selling. Soon, people from other parts of the coast were asking if I could make maps for their own beaches. My website 'Tidal Compass' was born.

It was January 2014 and my daughter Ottilie was steadily making my partner Naomi's belly swell in size. A few months before her arrival date, I devised a plan to sail around Britain, making new maps for the coastlines we explored. Sensibly, we decided to tone down the adventure and travelled in a camper van instead. With mobile wi-fi and off-the-grid power solutions, I was able to run my online shop from a laptop in the camper. If I could make this my business, we would in a sense live off the tides. I'm also a keen surfer, so I can't pretend that this wasn't also a brilliant way of combining waking up somewhere new almost every day, expanding my family's horizons in a very real way and getting out into the water myself to do something I love.

As we discovered new parts of Britain I noticed other flows of water that were leaving people equally baffled: whirlpools in Scotland, rips in Cornwall, rapids in Wales. And I realised how each of these phenomena has shaped the lives of all those who interact with them. Gradually I learned how we could try to engage with them.

Using my maps and graphics, this book dedicates a chapter to each of the eight main motions of water that affect Britain, exploring what makes water move in that way and how if affects adventures along the coast. *The Book of Tides* is for those who want to understand better how the waters of nature affects our daily lives and how it imperceptibly, but crucially, shapes both our actions and has shaped our landscape. It's for anyone who knows and loves our coast and who wants to understand, discover, surf or sail it better.

William Thomson
January 2016

When I first learnt about stream, I had a special way of remembering what time water flowed in which direction. I would picture a tide clock with high tide at the top and around that would be a compass face with north at the top beside high tide. Because these two were together I never forgot that water flows north at high tide. The next bit to remember was the times of slack water and I would also visualise these on the tide clock face.

For more than a year, this picture remained firmly in my mind until I was talking to a neighbour about the best times for her kids to go kayaking. She was worried that at certain tides they would be swept out to sea, but I explained how they would only be swept along the beach and, if they timed their adventures for slack water, they could easily stay in one place opposite their beachfront house. To help explain the cycle of stream, I transferred my tidal compass picture onto paper for the first time.

The first sketches were very basic, but I soon developed the design and put them in a local shop, Dunlin and Diver. Opposite is a photograph of a framed Deal map which they stocked. But how does it work? Starting from the middle and working out, the central circle shows the coast that the map represents. The small white ring with black and grey arrows inside shows the position where the information is specific, as well as the main directions of flow [north and south]. Alongside the central circle is the tide clock face and for the hours written in black, water flows in the direction of the black arrow. For the hours written in grey, water flows in the direction of the grey arrow. Outside of this are the black and grey rings showing the relative speeds of flow. Where the black and grey meet is slack water. The rings then show 'speeding up', 'maximum flow' and 'slowing down'. As an example, maximum flow north is around 1 hour after high tide, while maximum flow south is around 5 hours before high tide. The toggles at the top and bottom are for quick reference as to what direction water is flowing. In this book you will also see another style of map – the circle. This doesn't have the outer toggles so the only arrows showing the direction of flow are within the small white ring in the middle.

Using the information from my maps, Naomi and I would embark on adventures along the coast on an old 12"-windsurf board I had converted into a paddleboard. We would time our trips to drift with the stream until it turned and took us back home. While I was the paddler [I didn't actually need to paddle] Naomi would sit on the back and reel in the mackerel. If that's not living in harmony with nature, what is?

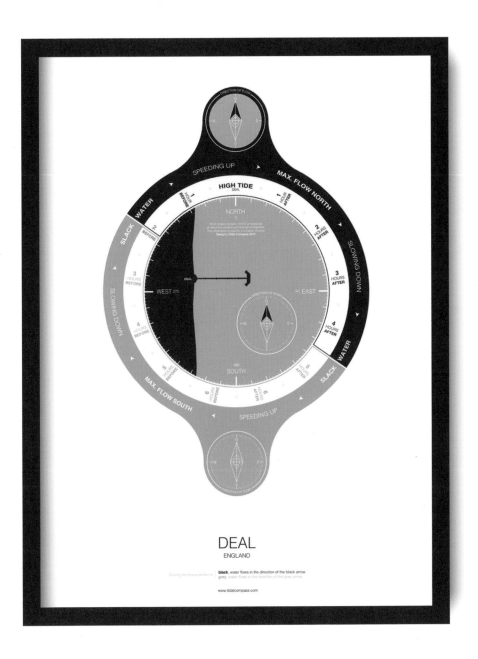

DEAL
ENGLAND

During the hours written in **black**, water flows in the direction of the black arrow
grey, water flows in the direction of the grey arrow

www.tidalcompass.com

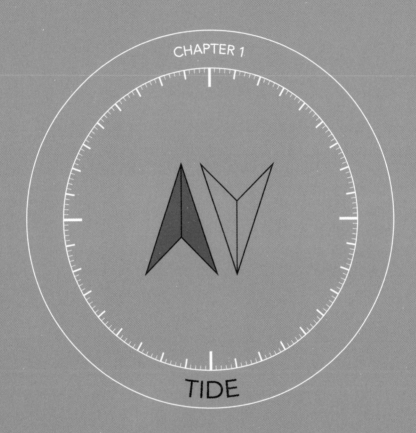

CHAPTER 1

TIDE

__tide__ the vertical motion of water

Tide is the vertical motion of water – so subtle that it is impossible to see with the naked eye. In Lyme Regis, I attempted to disprove this theory by putting aside the day to watch the tide drop [a very relaxing scientific experiment]. As the hours passed, I became aware of a general lowering of water against the harbour walls, but disappointingly failed to see a single vertical motion of water. However, we set the camera on time lapse and it clicked away a photograph every thirty seconds. On returning to the camper, we replayed the footage in fast forward and watched in fascination as water drained from the harbour until the boats were lying haphazardly in the mud.

As the earth rotates on its axis, the changing gravitational pull from the moon powers two giant waves flowing around the coast of Britain. The distances between peaks and troughs of the waves are roughly 580 kilometres and when a peak of a wave passes a beach it is high tide and when a trough passes it is low tide. It takes roughly 6 hours, 12 ½ minutes for a trough to reach a beach after a peak has passed, and this is the time between high and low tides.

The two waves begin their British adventure at Lands End. One travels north up the west coast, around the tip of Scotland, then down the east coast. The other flows up the English Channel and the two converge off the Thames Estuary. When a peak is at Lands End, it is also high tide around the west coast of Scotland and Yorkshire. At the same time, there are troughs in north Wales, north-east Scotland and the Thames Estuary.

For a small island, Britain is packed with a rich variety of coastal landscapes and the tide wave interacts with each one differently. In the tidal flats of Morecambe Bay, the effect of the tide is completely different to the marshes of Norfolk or the tidal River Thames and in this chapter we're going to explore the unique risks and rewards that ebb and flow with the tide.

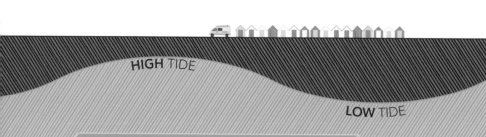

HIGH TIDE

LOW TIDE

TIDE **WAVE** TRAVELS **ALONG THE COAST**

As I discovered in Lyme Regis, it's extremely difficult to spot tide in just a few minutes. However, if you find yourself on a new beach without access to a tide table or tide app, here are some tips for quickly analysing the state of the tide.

Look out for a line of detritus. The high tide often leaves a clearly visible line of driftwood, shells, seaweed and [tragically] plastic. Among our travels around Britain, we also discovered a sheep, a seal, a porpoise and a fox – all in varying states of decomposition.

Look at sand texture. If the beach is sandy, the sand above the high tide line will be rough from footsteps, while the sand in the intertidal zone will be washed smooth.

Watch people. If there are fishermen around take note of where they have pitched their tents. Judging by their lethargic sport, it is unlikely they are constantly moving all their equipment as the tide rises.

Look at vertical structures. If you are somewhere there are piers, harbours, seawalls or cliffs – look out for marine organisms and seaweeds. The high tide line is the highest point these grow at and the surface above is usually lighter. Above the high-tide line you will find Rock Samphire [it doesn't like to get its feet wet].

Is the ground wet on a dry day? This is a clear sign the sea has just been on the beach where you are now standing. This is more notable in the morning before the heat from the sun dries the beach.

What way is water flowing? If you are near a tidal river or estuary, look out for the direction water is flowing [see page 48 – How to spot stream]. On a rising tide, water will be flowing inland and on a falling tide, water will be draining out to sea.

Gasping fish? The chances of this happening to you are extremely low, but it's such a clear indication of an ebbing tide that I've got to mention it. I was once walking beneath the White Cliffs of Dover when I stumbled upon a huge gasping sea bass. It's highly unlikely it leapt the three metres from the water's edge – it would have been caught out by the falling tide.

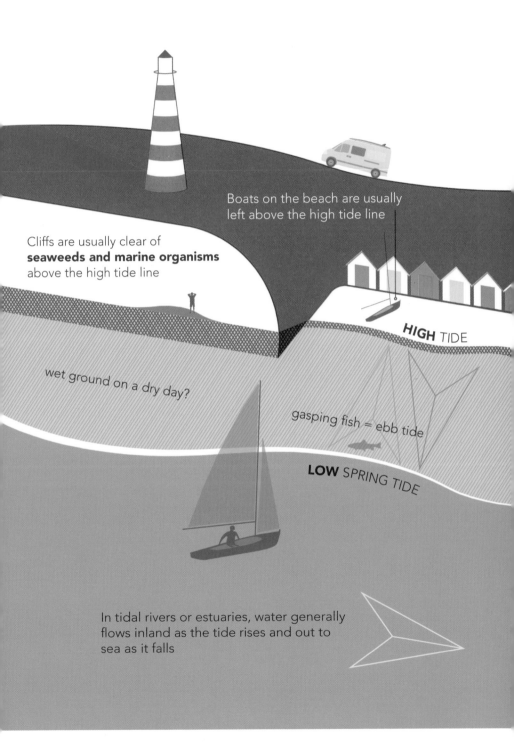

If you can see the moon in the day you can calculate the time of high tide from the phase of the moon

Boats on the beach are usually left above the high tide line

Cliffs are usually clear of **seaweeds and marine organisms** above the high tide line

HIGH TIDE

wet ground on a dry day?

gasping fish = ebb tide

LOW SPRING TIDE

In tidal rivers or estuaries, water generally flows inland as the tide rises and out to sea as it falls

The key theoretical timings in a semi-diurnal tidal cycle [the most common around Britain] are 6 hours 12 ½ minutes from high to low, 12 hours 25 minutes from high to high and 24 hours 50 minutes for a full cycle. The hours represent quarter, half and full rotations of the earth while the minutes are linked to the simultaneous orbit of the moon. In order to explore what this is all about we must take a step back and look at Britain on a global level.

Firstly, what makes the tide change? If you draw a line through the centres of earth and moon [see diagram opposite] the positions A and C experience the strongest gravitational pull, forming high tides where the sea bulges out from the seabed. At the same time the pull is weakest at B and D, resulting in low tide. If you imagine the bulges remain in position while the earth spins on its axis the effect is a giant wave with two peaks and two troughs flowing around the world.

If Britain is at position A at midday, it would rotate around to B in six hours and move from high to low tides. At midnight, it would arrive at C and high tide again. At 6am, it would pass another low tide and midday the next day would be high tide once more. This would be the case if the moon did not move, but in the time it has taken for Britain to arrive back at A the moon has moved around to position L. It takes another 50 minutes to realign with the moon at 'a' and this is why a tidal cycle is 24 hours, 50 minutes long. Following this theory, it takes 25 minutes to realign after a half rotation which is why there are 12 hours, 25 minutes between high tides.

Although the timings are consistent in a semi-diurnal cycle [but not precise due to many other factors] there is one main problem with the concept – land. If the world was pure ocean with a consistently smooth seabed, the tidal wave would flow just like this. However, irregular coastlines and seabeds break up the journeys of the tide waves and each continent or island has its own unique waves. As we discovered on the last page, Britain has two main waves that split when they reach Lands End and meet again north of the Thames Estuary. Each wave has its own unique quirks, created by the coastline and seabed it encounters on its journey around this varied island.

A typical tidal cycle in Britain takes **24 hours, 50 minutes**.

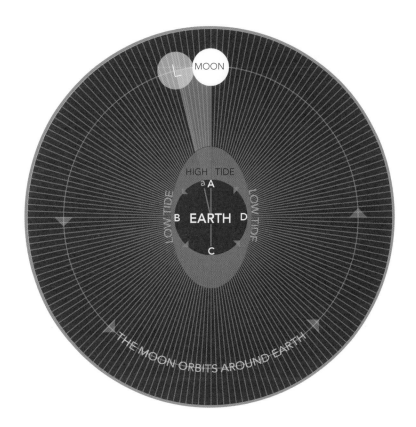

MOON

HIGH TIDE

a A

LOW TIDE
B **EARTH** D
LOW TIDE

C

THE MOON ORBITS AROUND EARTH

SUN

While the moon has the strongest effect on daily tides, the gravitational pull from the sun [one-third the power of the moon] plays an important role in monthly tides. However this is dependent upon where the moon is in its 29.7-day orbit of earth.

When the moon is aligned with the sun and earth the combined gravitational pull of the moon and sun is stronger, causing a higher tidal range with more pronounced highs and lows. These are known as 'spring' tides and happen twice a month, just after full and new moons. Spring tides have nothing to do with the season of spring but mean 'to spring forth' with speed.

A week after spring [and a week before], we have neap tides meaning 'without power'. These happen when the moon is perpendicular to the sun and earth so the combined pull is weaker. Neaps have a lower tidal range with less pronounced highs and lows.

Looking at a tide graph you can see whether you are going from springs to neaps or neaps to springs. Look at the heights of highs and lows over a few days. If they are getting closer each day you are coming off springs, and if the tidal range is increasing you are getting closer to springs. You can apply this theory to a tide table on a daily basis. If the second high tide is higher than the first you are getting closer to springs – if it is lower you are coming off springs.

My favourite aspect of tides is their ability to essentially 'reset' themselves. Because a daily tidal cycle is 24 hours, 50 minutes, high tide is 50 minutes later each day. Over a week this adds 6 hours. This means that every spring and neap high tide is the around the same time. It's easy to remember in my hometown of Deal because springs high tide is around 12am/pm and on neaps high water is around 6am/pm [give or take 50 minutes]. Try this with your local beach – the timings will be different but the concept is the same. With this knowledge you can even work out the time of high tide by knowing the phase of the moon.

In Britain **springs** occur **1–1.5 days after** full and new moons

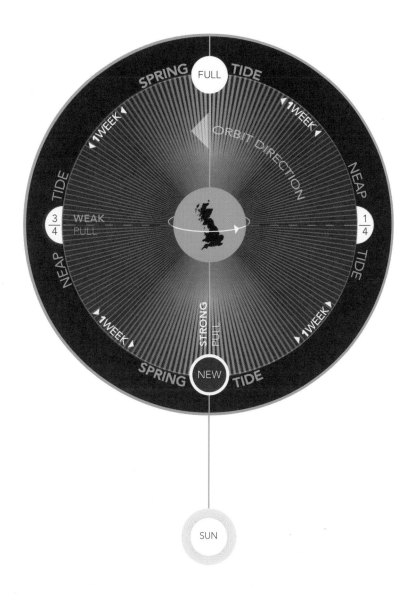

Pembrokeshire has some of the most rugged cliffs in Britain. The stretch from Fishguard to St Davids [Britain's smallest city] is particularly precipitous and the almost 300km Pembrokeshire Coastal Path will take you high above the sea. However, we're more interested in exploring the waters edge far below.

Low tide uncovers edible treats for supper. In Abercastle, Naomi and Ottilie foraged for crustaceans, seaweeds and molluscs while I was distracted by a plaque about Alfred Johnson – the first person to sail single-handed across the Atlantic. In August 1876 Johnson sailed into this cove after almost 60 days at sea in a fishing dory [essentially an open top dinghy]. He was in such a poor state the locals had to carry him from the boat. When asked what inspired such a foolhardy mission, he replied, 'because I'm a damned fool, just like they said!'

Hidden amongst the pebbles and rocks of Pembrokeshire are the 470-million year old fossilised remains of a colony of tiny sea creatures – the graptolite. In search of this fossil that looks like a minute hacksaw blade, remember to keep an eye on the tide and make sure you know whether it is coming in or going out. The biggest danger within this landscape is being cut off by the rising tide and lifeboat crews around Britain are constantly rescuing people from beneath cliffs.

CLIFFS ENVIRONMENT

At **high tide,** there is a danger of being **cut-off** beneath cliffs

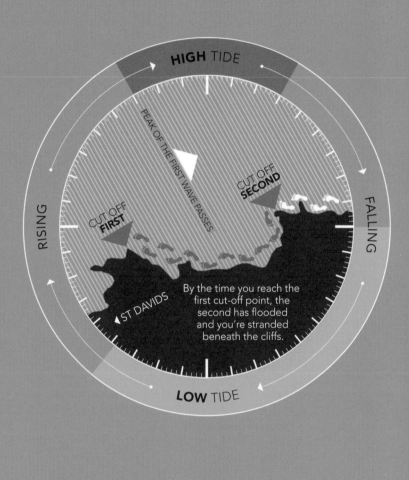

HIGH TIDE

PEAK OF THE FIRST WAVE PASSES

CUT OFF
SECOND

CUT OFF
FIRST

RISING

FALLING

ST DAVIDS

By the time you reach the
first cut-off point, the
second has flooded
and you're stranded
beneath the cliffs.

LOW TIDE

At **low tide** fossils and edible **treasures** are exposed

The great thing about climbing a few metres above the sea is you're guaranteed a soft landing if you fall off [make sure the water is deep enough first and remember the water depth at low tide can be several metres shallower than at high tide]. Sometimes you will have to traverse along a cliff face if the tide has cut off the lower beach.

Many hard rocky landscapes have been weathered smooth by waterfalls and blowholes. These make perfect natural slides for adults!

Many caves are filled with water at high tide, but the falling sea will allow you to explore as far as you dare. A waterproof torch is a handy piece of kit here.

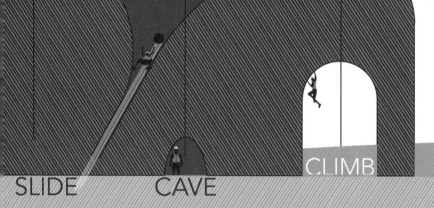

SLIDE   CAVE   CLIMB

Coasteering was pioneered by surfers in Wales in the 1980s. On waveless days the coastal explorers would scramble along rocky ledges, jump off cliffs and swim deep into caves. Now, adventure companies take people of all thrill-seeking, skill and fitness levels out along the coast.

A helmet will protect you against falling rocks from above as well as from any bangs to the head you may incur while coasteering.

A wetsuit and buoyancy aid will keep you warm and floaty. The idea of a wetsuit is to trap a thin layer of water that warms up between your skin and the suit.

Rubber boots protect your feet when climbing over sharp rocks. Wetsuit boots will also keep your feet warm.

You don't actually need to be able to swim when coasteering because the wetsuit and buoyancy aid will make you naturally float. Simply kick your legs.

Beat your personal jumping record by finding a cliff with multi-height ledges and work your way up. Always check water depth first.

SWIM

JUMP

Coasteering is a fantastic way to explore the geology and wildlife of the intertidal zone at close quarters. Although it can be a high-adrenaline activity, you can tone down the risk for any skill or fitness level by choosing more sheltered areas of the coast to explore, avoiding caves and only jumping off low ledges.

There are many edible plants on the coast and one that doesn't like to get its feet wet is rock samphire. Always just take a few cuttings from each plant to ensure conservation.

Crabs hide in mini caves when the tide falls, patiently waiting for the next high tide. Gently prod your hook into a likely looking cave and hopefully a crab will grab it.

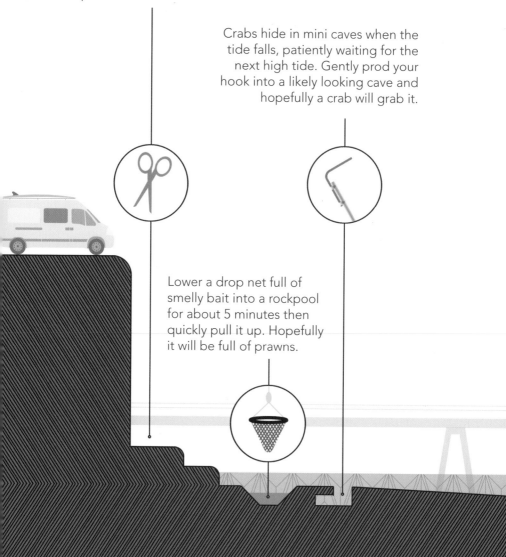

Lower a drop net full of smelly bait into a rockpool for about 5 minutes then quickly pull it up. Hopefully it will be full of prawns.

People have been foraging the British coast for as long as there have been humans here – thousands of years. Special flavours go in and out of fashion such as oysters which used to be a food for the poor and now are an expensive delicacy. Follow your taste buds, not the fashion!

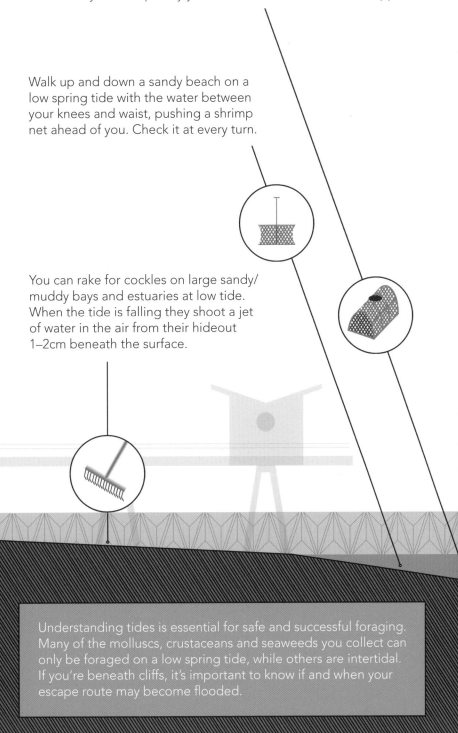

Wade out waist deep on a low spring tide and leave a lobster pot among the rocks. Return every low tide for the next couple of days and hopefully you will have a fresh lobster for supper.

Walk up and down a sandy beach on a low spring tide with the water between your knees and waist, pushing a shrimp net ahead of you. Check it at every turn.

You can rake for cockles on large sandy/muddy bays and estuaries at low tide. When the tide is falling they shoot a jet of water in the air from their hideout 1–2cm beneath the surface.

Understanding tides is essential for safe and successful foraging. Many of the molluscs, crustaceans and seaweeds you collect can only be foraged on a low spring tide, while others are intertidal. If you're beneath cliffs, it's important to know if and when your escape route may become flooded.

For a change of scenery we popped into London for some city culture. Looking around at the suits and ties, I wondered why nobody was dressed in swimwear and beach towels – don't they know at spring tides an expanse of golden sand runs from the London Eye all the way along to the Tate Modern?

We followed the beach all the way to the gallery. Within the peaceful space of the converted power station it's easy to forget a rising tide out at sea is flooding up the river as far as Teddington Lock, 89 kilometres from the estuary where river meets sea. By 3pm the Thames had risen by 7m and a brackish mix of salt and fresh water was lapping against the reinforced riverbanks.

The tide can get even higher. When low air pressure conspires with an easterly wind and spring tides, the river can reach alarming levels. Combine this with the post-glacial tilting of Britain in which the south-east is sinking into the sea at 5cm per century, and London is in serious risk of flooding. To protect the capital the Thames Flood Barrier was opened in 1982, and on especially high tides the gates can be closed to prevent any more water flowing upstream. Once the peak of the tide wave has passed the gates can be re-opened to allow the river to drain back out to sea.

Once the threat of flooding has subsided, Londoners can relax. On spring tides, low tide is always around 9pm – the perfect time for a beach party. Once a month, when the bright lights of the city are dazzled by the natural glow of the full moon, partygoers flock to the Southbank to enjoy music, sand sculptures and fire dancers. However, there's a strict time limit to the fun: high tide.

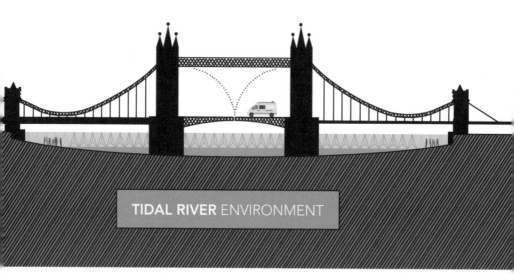

TIDAL RIVER ENVIRONMENT

At **high tide,** there is a risk of **flooding** in the city of London

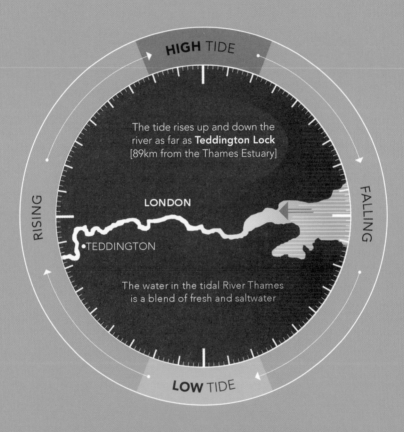

HIGH TIDE

RISING

FALLING

The tide rises up and down the river as far as **Teddington Lock** [89km from the Thames Estuary]

LONDON

•TEDDINGTON

The water in the tidal River Thames is a blend of fresh and saltwater

LOW TIDE

Sunbathe or **party** on a river beach at **low tide**

High and low tides are not always the height as stated on the tide table. The peaks and troughs of the tide waves are predicted by the positions of the earth, moon and sun in relation to each other. However, the weather conditions of the day can have a profound impact on the tide. The two main variables are air pressure and wind.

A low-air pressure system and onshore winds will conspire to create exceptionally high tides. Low air pressure occurs when warm air rises, relieving the pressure on the surface of the sea. A single millibar drop in air pressure can raise the local sea level by 10mm. How can you spot a low-pressure system? It's simple – the weather is atrocious! As the air rises it cools, forming into clouds that lead to wet and windy conditions. When we were living in a sixteenth-century smugglers' cottage on Deal seafront, a low-air pressure system was threatening a storm surge and flooding of low lying coastal areas. We were in direct threat so headed to the beach to fill-up some sandbags. I will never forget being thrashed by wind, hail, lightning and thunder as we shovelled wet sand into hessian sacks. In the end, we didn't even need them because although the tide rose perilously high, the offshore wind held it at bay. If there had been an onshore wind, the sea would have been blown further up the beach and into town.

To read about the extreme cases of low air pressure resulting in storm surges, go to page 122.

A high-air pressure system and offshore winds bring lower tides than expected. High pressure occurs when cold air sinks and exerts greater force on the surface of the sea, preventing the tide rising to its normal height. The weather conditions of a high-air pressure system are clear blue skies and low winds – perfect for exploring the beach. If the high-air pressure system occurs during springs, the sea will fall to unusually low levels. This is a perfect time to head down to the water's edge and search for hidden treasures that are normally submerged. If the wind is blowing from the land out to sea [offshore] it will hold back the incoming tide, giving you even more time to explore.

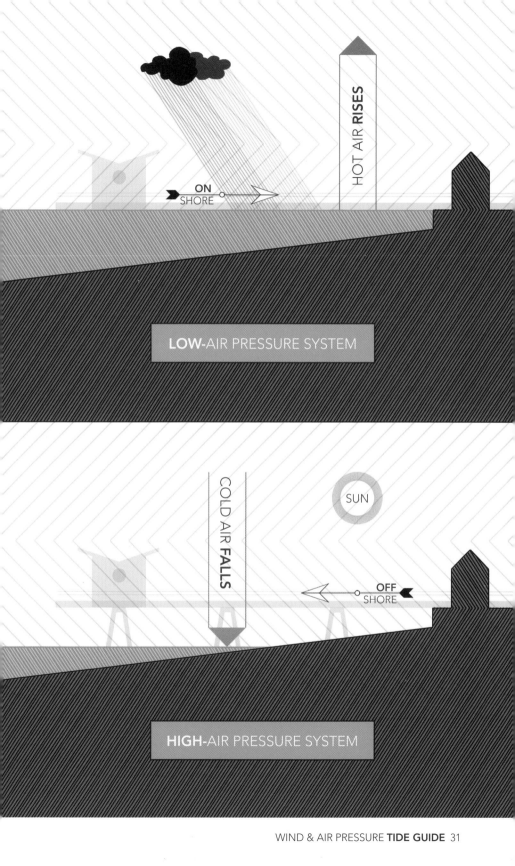

HOT AIR **RISES**

ON
SHORE

LOW-AIR PRESSURE SYSTEM

COLD AIR **FALLS**

SUN

OFF
SHORE

HIGH-AIR PRESSURE SYSTEM

We visited Morecambe Bay on a Saturday – campsite day. We took this opportunity to stock up with water and supplies then for the rest of the week we would live wild. The campsite, overlooking Britain's largest tidal flats, was similar to others around the island except for one detail: almost every tent had a three-wheeled buggy parked outside.

Sitting on a hill overlooking the bay we watched the effect of the tide wave's peak passing to the north and revealing 310sq/km of sand. The purpose of the buggies soon became apparent when the wind picked up and a dozen of the contraptions raced across the vast bay at speeds of up to 112kmh – powered entirely by a kite blowing in the wind.

It's not just kite buggies that enjoy the low-tide playground. Horse riders, bird watchers, dog walkers and cockle pickers all share the bay. The cockle pickers venture the furthest from land, to banks several kilometres from the high-tide line. This far out, an incoming tide can present a deadly danger. This was tragically highlighted in 2004 when 23 cockle pickers died from drowning and hypothermia when they were cut off by the incoming tide. Because the bay is so shallow, a small increase in the tide floods a huge area in very little time. To keep people safe a horn echoes across the landscape to herald the arrival of the next peak of the tide wave – making a sensible time to return to the high tide line.

With the tide racing in faster than a galloping horse, and rising to 10m above the sand, this is not the time to get stuck in Morecambe Bay's quicksand. Can you imagine being stuck waist deep while the incoming waters rise up your chest? The best way to avoid getting stuck is to explore the bay with Cedric Robinson, the royally appointed 'Queen's Guide to the Sands', but if that's not possible then turn to the next page to learn how to get out of quicksand.

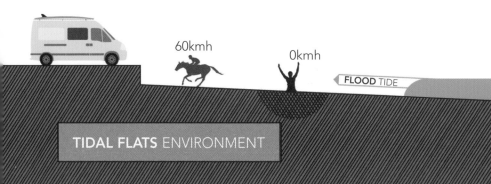

60kmh

0kmh

FLOOD TIDE

TIDAL FLATS ENVIRONMENT

The **rising tide** has killed people on Morecambe Bay. Beware.

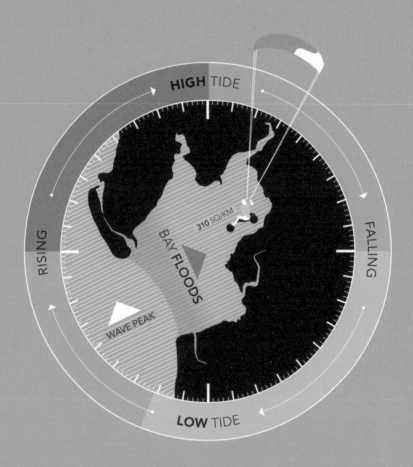

HIGH TIDE

RISING

FALLING

310 SQ/KM

BAY FLOODS

WAVE PEAK

LOW TIDE

A **falling tide** is the time to explore the bay – avoid quicksand!

Quicksand alone is not dangerous. The real threat is being stuck on a rising tide. For people who panic, their frantic movements sink them deeper into the trap while the flood tide rises around and above them. Don't worry – this nightmare will never happen to you because by the end of this guide you will know exactly how to get out before the tide gets up.

Let us start with some facts. What is quicksand? It is simply normal sand that has become so saturated in water that the friction between the particles is reduced and it turns into a viscous material that cannot support weight. When a moving object is placed on top, the vibrations liquefy the substance and the object will sink. Luckily for humans, we are half as dense as quicksand [1 gram per millilitre as opposed to 2 grams per millilitre] so will naturally float. If you stay calm, you will never sink further than your waist. If you become stuck with a backpack on, take it off because it increases your density and makes you sink further.

The best form of defence is prevention. Don't get stuck in the first place! Walk with a stick and use it to test wet sand before you walk on it. If you do feel yourself in a sticky patch, then take quick light-footed steps back. It takes a few moments for the quicksand to liquefy and if you are lucky you can get out before this happens. Walking barefoot also helps because hard soled boots increase the suction and hold you down. If these prevention strategies don't work, here are two simple steps for a safe escape:

1. Lie on your back and take a deep breath. This spreads your weight evenly across the sand, reducing the pressure from your feet and preventing them sinking deeper. Breathing deeply not only increases your natural buoyancy, it keeps you calm. Panicked movements create vibrations that liquefy the sand even more and make you sink deeper.

2. Swim backwards in slow motion. Slow movements reduce the vibrations, save energy and keep you calm. Wiggle your legs gently and sweep your arms in a long arc to propel you backwards as though you are swimming. When you reach dry sand, roll over and you are safe!

Quicksand is more dense than the human body so **you will naturally float**

FLOOD TIDE

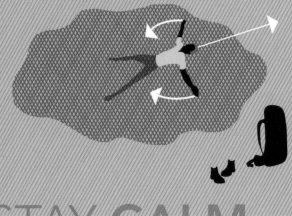

STAY **CALM**

If you are wearing a **backpack**, take it off immediately as it increases your density and makes your sink. If you can get your **boots** off, do so, because the hard soles increase the suction effect.

In a three-wheeled **kite buggy** you can travel at speeds of up to 100kmh and jump into the air. By steering the front wheel with your feet your hands are free to control the kite allowing you to speed up, slow down and even travel upwind.

Mounting a windsurfing mast onto the front of a specially built board will create a **land windsurfing** board. This is a great alternative to windsurfing in the winter when you don't fancy the cold water.

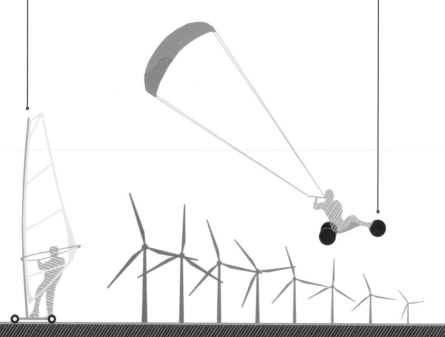

The concept of propelling a wheeled contraption by a sail or kite has been experimented with for hundreds of years with travellers commenting on Chinese designs in the sixteenth century. Serious racing of sand yachts boomed in the 1950s with British and International bodies being established to organise specific classifications for competitions.

Apart from the obvious kite/sail and board/buggy – a helmet is the minimum safety equipment required due to the high speeds.

Kite landboarding is a blend of kite surfing and skating and allows you to reach top speeds of up to 60kmh and jump high into the air.

The classic **sand yacht** is the closest to sailing a dinghy and you can reach speeds of up to 112kmh.

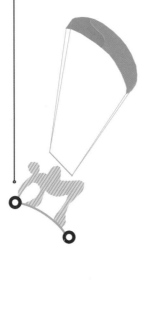

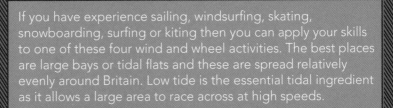

If you have experience sailing, windsurfing, skating, snowboarding, surfing or kiting then you can apply your skills to one of these four wind and wheel activities. The best places are large bays or tidal flats and these are spread relatively evenly around Britain. Low tide is the essential tidal ingredient as it allows a large area to race across at high speeds.

In north Norfolk, the beaches are cut off from the coastal villages by a network of tidal creeks and marshes. Occasionally a road bisects the marsh, allowing access to a desert island style expanse of white sand. In your excitement to get to the beach be aware of one detail – when a peak of the tide wave passes Norfolk, water floods through the narrow openings to the marshes and floods a vast network of creeks. When this happens, the roads turn into rivers and the car parks become lakes.

We took one of these roads through the marsh at low tide and parked on a raised bank above the car park. With my office facing an empty creek, I settled down for the morning work session. Every few minutes I would look up from the laptop to see the fishing boat in the creek floating a little higher. When the creeks were full enough the skipper arrived and negotiated his way through the tidal waterways towards his fishing grounds in the North Sea. The water level continued to rise until it was so high that corrosive saltwater spilled over the banks into the car park, pouring down the road towards civilisation. We became stranded in splendid isolation. Looking out the window, I noticed we were not quite alone. Below us, water was splashing against the tyres of a shiny sports car. The corrosive saltwater rose up the wheels and gently lapped against the engine.

High tide does not last long. As the tide wave continued its perpetual journey around Britain, the trough approached from Yorkshire heralding low tide. By the time the owner of the car returned, the car park had dried out, the creeks were empty and he couldn't work out why his seats were wet and the engine wouldn't start. Finally, it coughed into action and he disappeared down the road in a cloud of smoke.

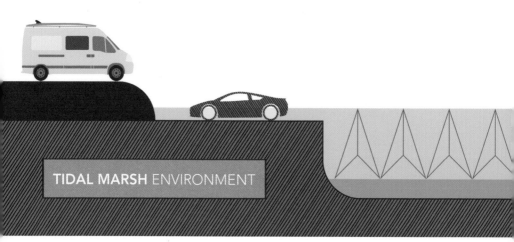

TIDAL MARSH ENVIRONMENT

On high springs, **car parks become lakes** and roads turn into rivers

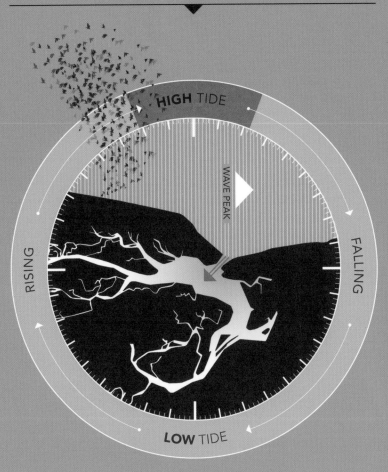

HIGH TIDE

WAVE PEAK

RISING

FALLING

LOW TIDE

Raised walkways allow access deep into the marsh at high tide

Dunes are a good vantage point to watch both the intertidal zone and the marshes. They also support important habitats for species such as butterflies and dragonflies.

High tide forces oystercatchers, grey plovers, golden plovers, knot, dunlin, bar-tailed godwits and curlew off the intertidal zone and up into the air.

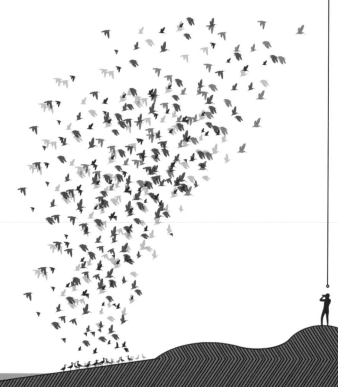

Birdwatching as a hobby started in the Victorian era with the establishment of the Royal Society for Protection of Birds [RSPB] in 1889. Before then birds were located mainly to be hunted instead of conserved. Now, there are three-million bird watchers around Britain and you can join in by letting the RSPB know which birds you see [or hear] so they can build an ornithological map of Britain.

Binoculars will help spot birds from far away and a **notebook** is useful for writing down your observations from the adventure.

Deciduous trees and bushes provide plenty of food for the resident birds, summer visiting breeders, passage migrants and vagrants.

Raised walkways allow you to access deep into the marshes at all tides. Look out for tired migrants taking refuge in isolated bushes during spring and autumn.

Hides are strategically placed to optimise viewing of waders and wildfowl, as well as ethereal barn owls hunting over the marshes.

The name birdwatching is misleading because listening is a key tool – you will hear many birds before you see them so it's a good idea to learn bird calls. An understanding of tides is essential in places like north Norfolk because many of the spectacles are at specific tides, such as thousands of waders flying up into the air as high tide cuts off the beach.

The Goodwin Sands are a group of tidal islands in the middle of the English Channel. At high tide they are completely submerged, but as the tide falls they appear to rise up from the sea. In true English fashion, a game of cricket was first played there in 1824, and visitors informally maintain the tradition on spring tides in the summer. The undulating landscape barely fits the requirements for a wicket, but cricketers are an eccentric bunch and the opportunity to play in the middle of the channel with a colony of grey seals for spectators is too good to resist. For people wanting more thrill than bat and ball, low tide exposes shallow lagoons between the islands to make a perfect kite-surfing arena.

Judging by their ancient name, 'ye olde shippe swallower', the sands are not all fun and games. In the days before hydrographic surveys and satnav, they were a treacherous death trap inconveniently located in the middle of the busiest shipping lane in the world. They were [and still are] especially dangerous on a mid-tide when the hard banks hide just below the surface. Without the aid of modern navigation crews would often find themselves run aground, especially in fog or on dark nights. Even sailors who knew the area well could become wrecked because strong streams [see next chapter] are constantly shifting the fine sand particles and reshaping the islands.

If a shipwrecked crew were lucky, the tide would be rising and they could attempt to float themselves off. If they were unfortunate enough to become wrecked on an uneven bank on a falling tide, there was a risk the ship would break up under its own weight. The sailors would then have a couple of hours to walk around the sands and come to terms with their impending doom before the flood tide engulfed their temporary island.

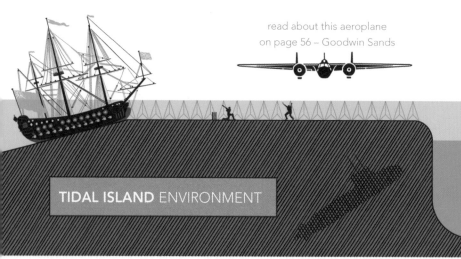

read about this aeroplane
on page 56 – Goodwin Sands

TIDAL ISLAND ENVIRONMENT

At high tide the **island disappears**

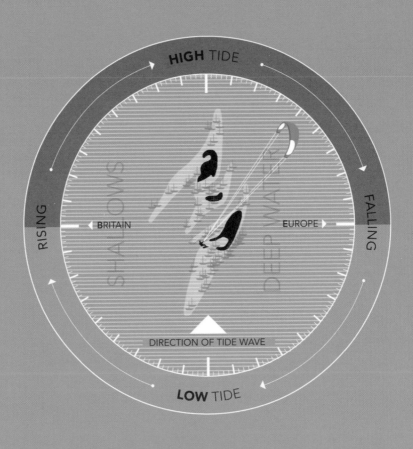

HIGH TIDE

RISING

SHALLOWS

BRITAIN

DEEP WATER

EUROPE

FALLING

DIRECTION OF TIDE WAVE

LOW TIDE

On a **low spring tide** you can walk on the sands

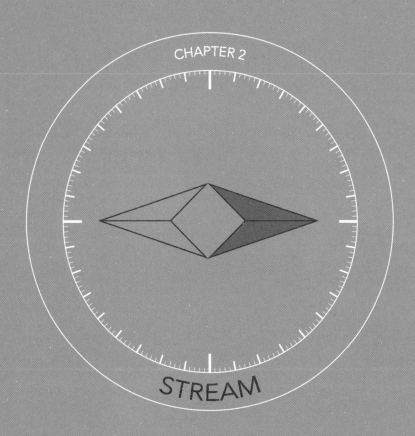

CHAPTER 2

STREAM

stream *the horizontal motion of water*

Stream is the movement of water back and forth along a coast. This flow of the sea is directly powered by the wave that makes tide, which is confusingly why stream is often called tide. In 1860, Sir William Thomson [no relation] cleared up the issue by stating 'sailors refer to the vertical motion of water in a harbour as tide and the horizontal motion of water at sea as tide'. This is a logical approach because the vertical tide is more important in a harbour, while the horizontal tide has more influence on your journey plans at sea. To keep this chapter simple, I'll be referring to the horizontal tide as 'stream'.

Because both motions of water [tide and stream] are made by the same wave flowing around Britain, it makes sense that they share certain qualities. But first, let us explore a typical stream cycle. In most places with a regular coastline and semi-diurnal tide, the stream will flow for just over 6 hours in either direction. Within each 6-hour period the flow will speed up for 3 hours, then slow down for the same time. The direction of stream changes when the flow is weakest and this is known as slack water.

Now you understand the unique properties of stream let's look at what links it to tide. The main connection is time – slack water is always at the same time before and after high tide. The time of slack water is unique to every beach around Britain, but through making my maps I have noticed that at high tide the stream is always flowing in the direction of the tide wave. On the west coast it flows north at high tide, on the east coast it flows south and on the south coast it flows east. In tidal rivers the stream usually flows upriver as the tide is rising, then flows out to sea when the tide is falling. It generally flows out for longer than it flows in.

Spring and neap tides have a huge impact on stream. During spring tides, when the tidal range is greatest, the strongest streams occur and slack water is very quick. Alternately, when the tidal range is reduced during neap tides the maximum flow is slower and slack water is more prolonged.

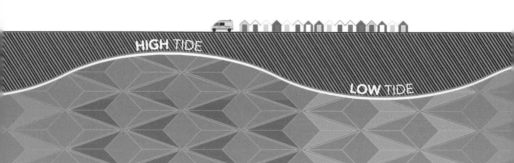

HIGH TIDE

LOW TIDE

On a calm day, it can look as though the sea is completely still. Do not be fooled! Water is always moving, even at slack water. Let me share a few tricks for spotting the direction of stream without getting your feet wet.

Look out for ships at anchor. The anchor chain is usually tied to the bow [front] of the boat and when there is no wind the boat will spin around on its anchor chain until it is facing the direction of flow. This is especially interesting at slack water because you can see a ship swinging around on its anchor as the direction of flow changes. If you watch this happen then you know the water will be flowing that direction for the next six hours [if the coast follows a semi-diurnal tidal stream cycle].

Look out for pier legs or posts in the water. When water flows past a stationary object it creates a disturbance downstream, just like when water flows past a rock in a river. By noting the side of the disturbance you can locate the direction water is flowing from and the greater the disturbance, the greater speed of flow.

Look out for buoys. You may be on a beach without ships at anchor or posts in the water, but the British coast is a busy fishing area and fishermen leave buoys as markers for where their pots are on the seabed. Because these buoys are secured to a stationary object the disturbance behind the buoy follows the same principle as pier legs. As buoys can be further out to sea, a pair of binoculars can come in handy.

Look out for fishing boats. Many cunning old seadogs will motor close to shore when they're heading into the stream because the flow is weaker in shallower water near the beach. But be aware they may simply be fishing close in!

Most beaches have driftwood on the high tide line – throw a piece in as far as you can and see which direction it drifts. You can even calculate the speed of flow by walking along with the driftwood and counting how many seconds it takes to cover a distance [speed = distance/time].

Look out for swimmers. Are they paddling vigorously and getting nowhere or are they casually racing past?

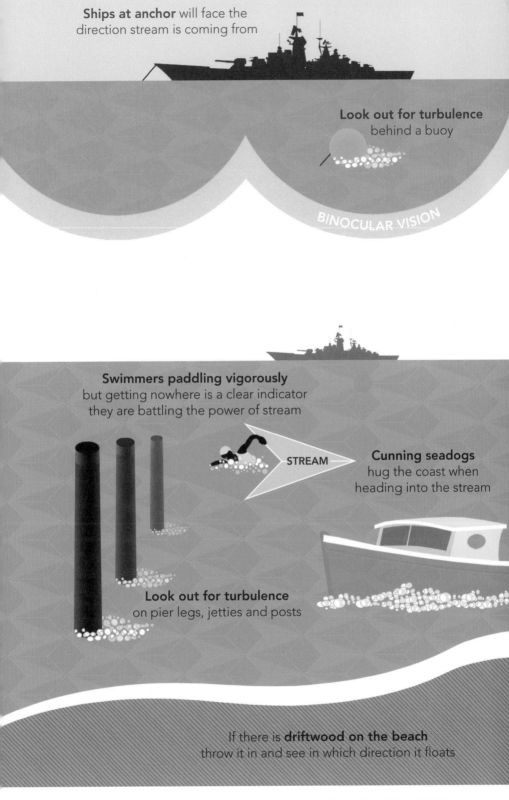

Ships at anchor will face the direction stream is coming from

Look out for turbulence behind a buoy

BINOCULAR VISION

Swimmers paddling vigorously but getting nowhere is a clear indicator they are battling the power of stream

STREAM

Cunning seadogs hug the coast when heading into the stream

Look out for turbulence on pier legs, jetties and posts

If there is **driftwood on the beach** throw it in and see in which direction it floats

Considering stream is one of the main flows of water around Britain, very few people have an understanding of its cycles. Lifeboat crews learn about it but there is still a murmur of 'what direction is the stream going?' Even the commodore of the local sailing club admitted to me that, although he remembers the times of slack water, he sometimes forgets which way the water flows at high tide. The people best acquainted with stream are the fishermen and our local lobster man 'Dave the Seadog' is no exception. His house faces his boat, which faces the sea. Naomi and I joke that Dave walks backwards to his house from the boat because we have never, ever seen him turn his back to the sea.

Unless you have lived your entire life on or by the water, like Dave, or benefitted from lifeboat training, how can you learn about stream? Before I designed my maps I used two main ways, but one is so mind-boggling that most people simply don't understand it, so I'll focus on the second.

A more visual format is a tidal stream atlas [illustrated on the opposite page]. Each box represents an hour before and after high tide and the arrows show the direction water is flowing. The bigger arrows represent faster flow and the smaller arrows weaker flow. That is why I designed my maps – to clearly show the cycle of speeding up, slowing down, changing direction and following the same pattern the opposite way. In this chapter, and throughout this book, you'll find many of my maps and the instructions are simple: for the hours written in black water flows in the direction of the black arrow, and for the hours written in white water flows in the direction of the white arrow.

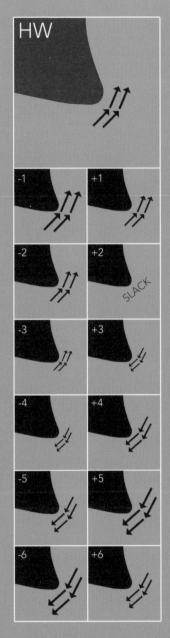

TIDAL STREAM ATLAS
[difficult to decipher cycle]

One of the most important streams affecting Britain is the Gulf Stream and although it has nothing to do with tidal stream [except for its river flowing qualities] it slots nicely into this chapter. The Gulf Stream is one of the strongest ocean currents in the world, transporting 150-million cubic metres of warm water per second from the Gulf of Mexico towards Britain and beyond. The channel can be 100km wide, 1km deep and reach speeds of 5mph. The Spanish were the first to discover the current in 1512 and harnessed its power when sailing from the Caribbean to Europe. It wasn't until Benjamin Franklin made a map in 1770 that the information was available in Britain, but embarrassingly British sea captains ignored him and didn't utilise the flow until many years later.

The current is created by two factors. The simple one is south-westerly trade winds literally blowing water towards Britain. The more complicated cause is the North Atlantic Deep Water [NADW] channel. Water in the North Atlantic becomes denser through cooling from Arctic winds and increased salinity [salt water content] as ice bergs form and discard the salt. This dense water sinks to the seabed and flows south to the equator. The purpose of the Gulf Stream is to transport water up to the North Atlantic to replace the equator-bound NADW.

Without the Gulf Stream, Britain would be 5° Celsius colder in the winter with December temperatures in London averaging a shivery 2°C. The effect is most noticeable on the west coast of Scotland where snow is rarely found at sea level on islands such as Skye. This doesn't stop the wind howling through the Isle – our camper felt like a yacht being tossed around a stormy sea as we headed up to the northern tip to look for basking sharks, killer whales [orcas], seals, porpoises and minke whales who are attracted here by the warm water and rich hunting. Sadly we didn't spot any marine mammals so retreated off the wet and windy shores, questioning the locals' word that 'winters aren't that bad here'.

With the importance of the Gulf Stream, it's worrying to hear that it is possibly slowing down as a direct consequence of global warming. As ice caps melt in the Arctic an increase in fresh water in the North Atlantic means there's less 'dense water' to join the NADW. With a weaker south current there is less need for a strong northerly current and some scientists predict the Gulf Stream will slow down 25% this century.

The last slow down of the Gulf Stream was 11,000 years ago when melting ice sheets in North America added fresh water to the Atlantic. With less 'dense water' in the NADW, the Gulf Stream slowed down with devastating consequences to Britain. In just a few decades, we slipped back into the ice age and although the effects this century won't be as catastrophic, we'll definitely need thicker wetsuits for winter adventures.

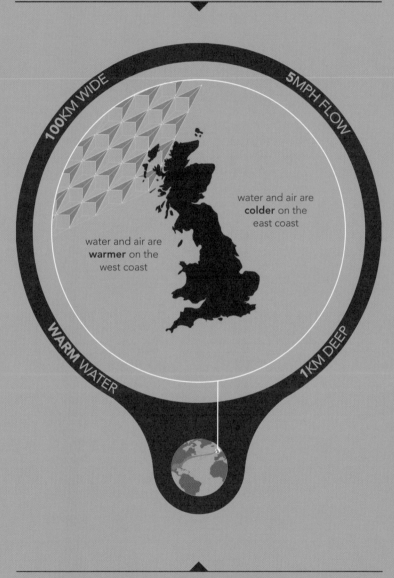

The Gulf Stream brings warm water to Britain from America

As we discovered in London, the tide can reach cities many miles inland and Bristol is no exception. The Bristol Channel boasts the second largest tidal range in the world [14 metres] and this is accompanied by strong streams reaching 8 knots. In the days of sail this was faster than the trading ships could travel so they were forced to plan their journeys to 'sail with the tide' when leaving harbour. As we learnt from Sir William Thomson's distinction, the word tide can also mean stream and in this case captains needed to time their departures for when the stream was flowing out to sea. This happens from ½ hour after high tide to six hours before the next high tide [50% of the time]. If you add the vertical tide into the equation it leaves only three hours out of every twelve when there is enough water depth to get out of the harbour, coupled with a favourable stream to get out to sea.

In the eighteenth century two thousand ships 'sailed with the tide' en route to West Africa to undertake the city's most shameful business – from 1730 to 1745 Bristol was Britain's main slaving port. Regardless of the time of day, ships would leave harbour within the tidal window three hours after high tide and head down to West Africa to trade British made goods for slaves. From West Africa the ships harnessed trade winds to sail across the Atlantic and then trade the men, women and children for American made goods to sell back in Britain. This route was known as the Slave Trade Triangle.

One of Bristol's prouder achievements was the development of the Bristol Channel Pilot Cutter, which is widely regarded as the greatest ever sailing boat. The vessel was able to sail against the stream so pilots could intercept large vessels at sea and guide them through the dangerous channel into harbour. The unique hull shape and sail configuration allowed a crew of two to make headway against the stream, while their shallow draught gave them a greater tidal window to negotiate the waters and harbours around the channel.

SLAVE SHIP

Water flows **downstream** from ½ hour after to 6 hours before

▼

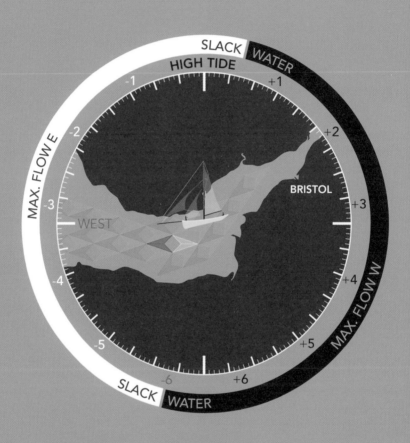

Water flows **upstream** from 6 hours before until ½ hour after

▲

On Monday 26th August 1943 – at the height of the Battle of Britain – a German Dornier 17 with the call sign 5K + AR became separated from its formation in clouds over the English Channel. An RAF Defiant took this opportunity to attack the bomber. With one engine disabled and the other damaged, the pilot Willi Effmert attempted a landing on the Goodwin Sands that were uncovered by the low tide 4,572m below. Just as the undulating sands make a challenging cricket wicket [see page 40] they make an appalling runway and the aeroplane somersaulted onto its back. Two of the crew died and their bodies washed up on different sides of the North Sea. How could this happen? A possible answer is that one body entered the water immediately and was taken by the southbound stream. If the tide was falling, the other body may not have entered the water until the flood tide several hours later by which time the two young men would have been many miles apart and subject to different streams.

The strong streams that washed the crew to opposing shores of the North Sea are constantly shifting the fine sand particles of the Goodwins. This phenomenon gave rise to their nickname 'ye olde ship swallower' because the sands wrap around wrecks until they are deep within the banks. This happened with the Dornier, a plane which lay uncovered for over 60 years until the same process washed away the sands and revealed her to scuba divers in 2008. A plan was devised to raise the only complete Dornier 17 in the world which we watched through a telescope on the Deal seafront.

The main challenge was tidal stream. Divers could only work at slack water and, as we now know, this happens for only two short periods every 12 hours. Once the stream builds up it stirs up the sand and reduces visibility to less than an arm's length. Added to this, it's almost impossible to undertake technical work against what's essentially a fast flowing river. The crew was hampered by bad weather during the slack water windows but, finally, the wind dropped at slack water at 5pm on 10 June 2013 and the historic plane was returned to the air once more.

Slack water is at 2 hours before and 4½ hours after

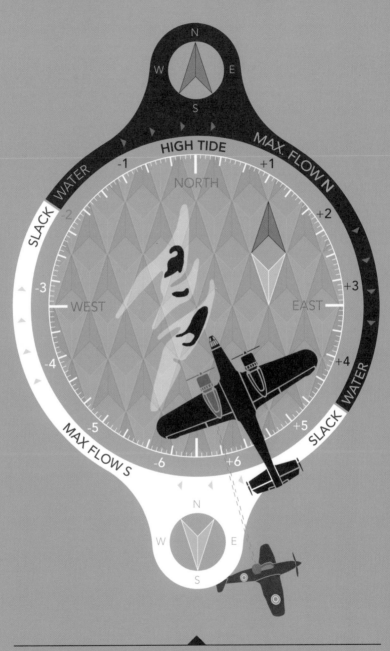

From 4½ hours after to 2 hours before, water flows **south**

Many of the best snorkelling sites around Britain are located in **sheltered bays** that are protected from the stream flowing along the coast. However, if you go around a headland be aware you will be exposed to the flow and also in an area where rip currents can develop.

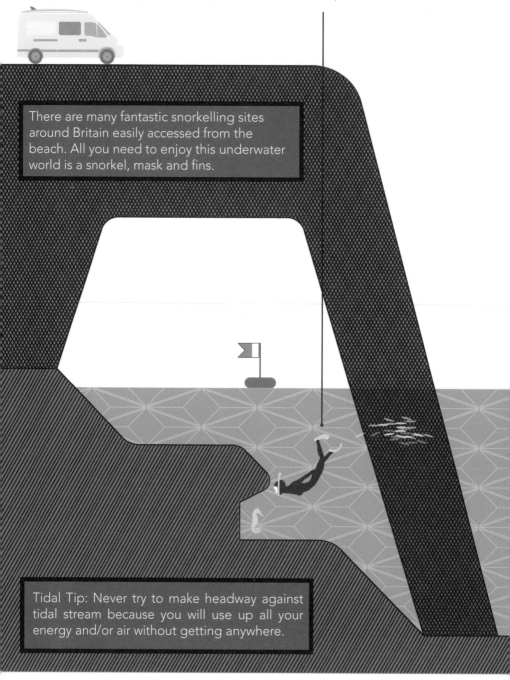

There are many fantastic snorkelling sites around Britain easily accessed from the beach. All you need to enjoy this underwater world is a snorkel, mask and fins.

Tidal Tip: Never try to make headway against tidal stream because you will use up all your energy and/or air without getting anywhere.

EQUIPMENT

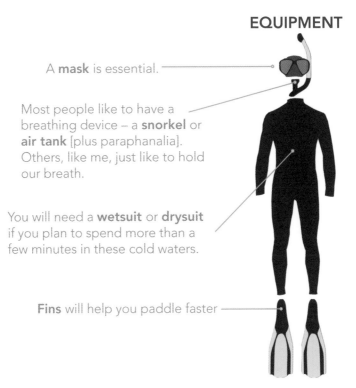

A **mask** is essential.

Most people like to have a breathing device – a **snorkel** or **air tank** [plus paraphanalia]. Others, like me, just like to hold our breath.

You will need a **wetsuit** or **drysuit** if you plan to spend more than a few minutes in these cold waters.

Fins will help you paddle faster

Unless you are in a bay sheltered from tidal flow then **slack water** is the only time to safely dive a wreck. If doing a drift dive, you can use the flow to take you along.

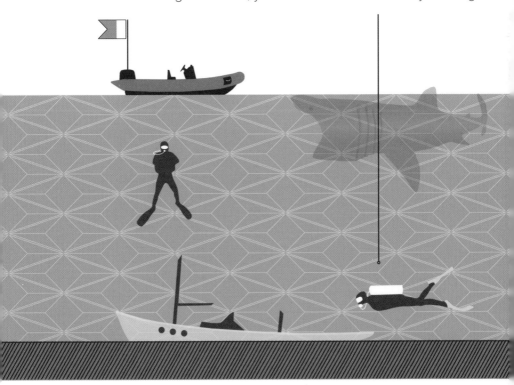

Every year since 1931 [excluding the World War II years when leisure boating was banned], sailors from Britain and overseas have flocked to the Isle of Wight to compete against each other and tidal stream in the Round-the-Island race. As the name suggests, the 50 nautical mile journey involves a complete circumnavigation of the island starting and ending in Cowes, the internationally recognised Home of Yachting since the founding of the Royal Yacht Squadron there in 1815.

The tidal patterns around the Isle of Wight are notoriously complex and each leg of the race experiences a different direction of stream. This is due to the geography of the Solent on which the northern half of the Island lies. One school of thought for the irregular tides is the narrow openings to the English Channel, while another theory is that the Solent is in the centre of the tidal wave that flows up the English Channel. Regardless of the cause, precise planning and preparation of the streams on the day can make the difference between fame and shame.

To simplify the challenges of tidal stream, the race is sensibly held at the same time before high tide Portsmouth each year, when the stream is beginning to flow west [the direction of the first leg down to the Needles]. Because the time of high tide is different every day, but can be predicted many years in advance, the most suitable Saturday in late May, June or early July is chosen when high tide Portsmouth is in the early morning. The race days are published five years in advance and the fascinating repetition of tides means we can predict the tidal stream for each leg of the journey for every hour of each race.

But it's not enough for skippers to simply know the direction of stream. To enhance the challenge, there are channels where the water flows faster and areas where the stream is weaker. There are even eddies where water flows in the opposite direction to the main stream. The secret to success for the 16,000 sailors in 1,700 boats is to position themselves in the most favourable channel regardless of whether they are sailing with or against the stream. So much for sailing being just about wind!

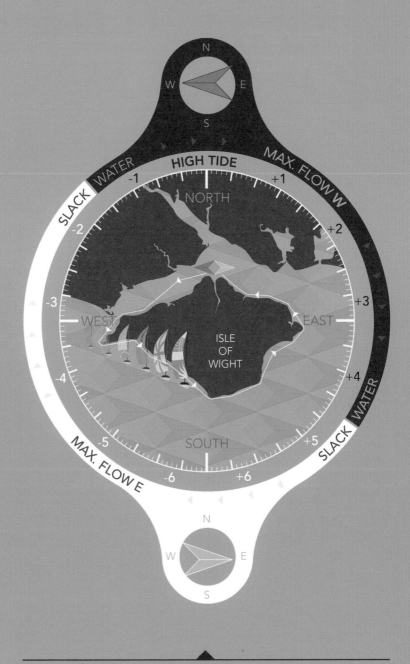

Suitable for cruising around Britain and crossing oceans, a **sailing yacht** allows you to harness the power of the wind to travel long distances. For windless days inboard engines allow travel against stream.

Powerful outboard engines allow you to travel fast in a **powerboat** and even tow a doughnut, waterski or wakeboard. Travelling into the stream will drastically increase your fuel consumption.

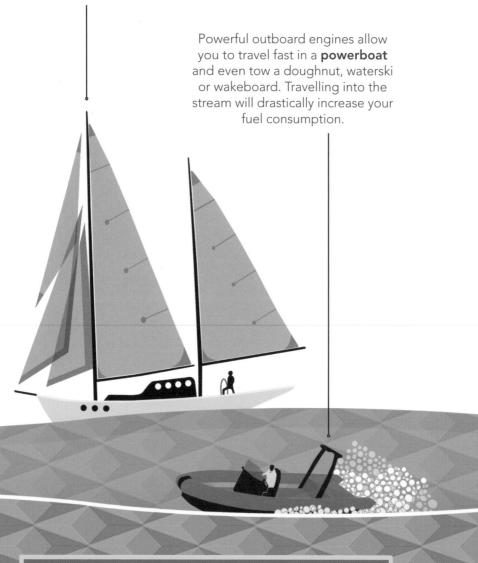

Boating in Britain has been about as long as humans. The original boats would have been hollowed out tree trunks and the last 8,000 years have seen huge developments in construction but the concept remains the same: to float and allow travel across water.

The ultimate in luxury is **mega yachting** and the multi-million pound price tags restrict them to the rich. Onboard features include helicopters, submarines, jacuzzis and various bars and lounges for entertaining guests. A knowledge of stream is not required for the owners as a full time crew takes care of the boating.

Dinghy sailing is mainly a competitive hobby around Britain although they're also fun for exploring the local waters. With only a sail for propulsion you're at the mercy of stream if/when the wind drops. If exploring a coast plan your return journey to go with the flow to compensate for this eventuality.

No matter how big and powerful your boat is, understanding the cycles of stream is essential for journey planning. In some places around Britain the stream reach speeds of double figures which will slow the travel of a high-powered boat and effectively halt the advance of a low-speed vessel.

Before we embarked on our voyage around the coast of Britain I worked on a high-speed rib [rigid hull inflatable boat]. It sounds hard to believe that this is actually a job, but people would pay us to take them at high speeds along the coast and out to the Goodwin Sands. The wetter our customers got the happier they would be, and I soon noticed a correlation between wind and stream that made some trips wilder than others.

In the summer the boat would run from dawn to dusk, during which time the stream would change direction at least once. When 'the tide turned' there would be a noticeable difference in the sea conditions – as long as the wind remained constant and was parallel to the coast. When the wind and stream were blowing and flowing in the same direction the sea would be flat. On these occasions it was difficult to get our passengers wet, but it did mean we could engage in adrenaline fuelled high-speed manoeuvres. As soon as the stream turned and started flowing against the direction of the wind, friction from the two opposing bodies of energy would create rough seas.

As the strength of stream sped up the roughness increased, most noticeably during spring tides when the maximum flow is at its fastest. On these occasions the boat would bump through choppy seas with huge clouds of white spray engulfing the deck every time we punched into a bank of water. During these trips we wouldn't be able to make sharp turns but we didn't need to – the ride was wild enough. These conditions would only last a short time because the sea would calm down as soon as 'the tide turned' and the stream flowed in the same direction as the wind.

This is particularly useful knowledge if you are stand-up paddle boarding or mackerel fishing (or both). The main disadvantage of paddle boarding is the difficulty to maintain balance in choppy conditions so it's worth planning your adventure when wind and stream are moving together. The sea state is also important when mackerel fishing because you want the water to be as calm and clear as possible, so the fish can easily see your shiny whitebait lure.

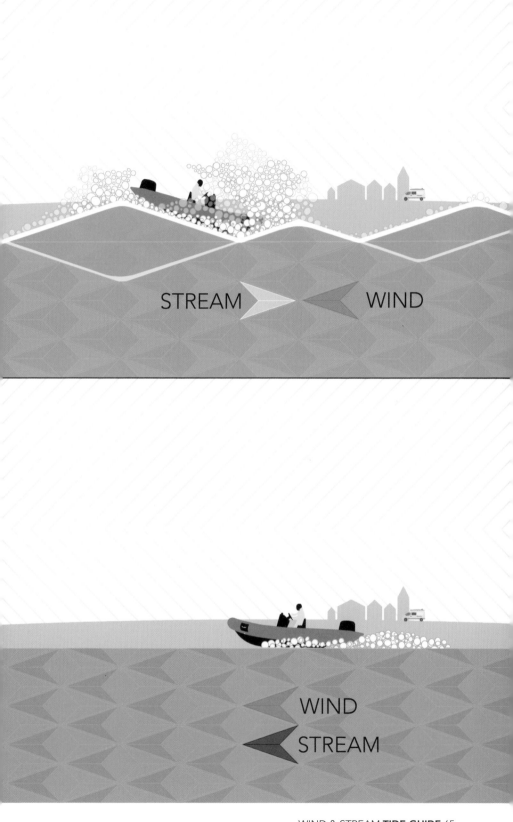

STREAM   WIND

WIND
STREAM

Paddleboarding is both **dynamic and theraputic**. The skill of balancing is blended with the relaxation of gliding over the surface of the sea and watching the underwater world below.

New technologies allow high performance **inflatable boards** capable of cruising the coast and surfing waves. When you get back to the car, it's just a case of deflating and throwing in the boot [carefully].

Getting used to the balance on a paddleboard takes a little time and **you've got to be prepared to get wet** when learning. When conditions get a little rough and the wind picks up even experienced paddlers can find it hard work.

When planning a paddling adventure always try to time your journey to **go with the flow**. Although paddling into stream is a great workout you won't get very far and as soon as you stop paddling to give your arms a rest the stream will drag you back to where you started.

Sit on top kayaks are the most accessible option for people of all ages and skill levels. In the rare scenario, they do tip over it's easy to get back on and you can even **mount a fishing rod on the back** to catch supper while you paddle.

Sea kayaks are great for **long distance adventures** [including overnight camping] because you can stow lots of gear on them and spend long days in the seat without getting too tired or uncomfortable.

You can reach **fast speeds** in a sea kayak and even make headway against the stream.

In the far north of Britain, there is a channel where some of the fastest tidal streams in the world can be found. This is the Pentland Firth, a narrow stretch of treacherous water between the mainland and Orkney Islands. Why is the flow stronger here than other parts of Britain? As the tide wave passes around the north of Scotland the water is funnelled through the Firth [which isn't actually a firth] and the ensuing build up in pressure generates streams of up to 5 metres per second [18 kmh].

These fast-flowing currents pose serious problems for shipping of all sizes so precise planning and preparation are necessary – as is battening down all the hatches. Yachts should always plan their crossing to go with the stream and once in the channel there's no going back (but lots of obstacles ahead). To name a few there are eddies downstream of all the islands, a whirlpool called the Swilkie and several tidal races [see pages 88-9]. The main consideration before entering the straits is the direction of wind and swell in relation to stream. While wind vs stream affects the sea state, up here swell from the North Sea and Atlantic is added into the equation. When the stream is running against swell the sea becomes far worse than wind against stream. To minimise these dangers it's recommended to wait for neap tides before crossing because the flow is weaker.

For the life-threatening challenges of the Pentland Firth there are also life-giving opportunities in the form of harnessing the tidal power to generate clean, green energy. These waters have been dubbed 'the Saudi Arabia of tidal power' with a potential to generate 20GW of electricity. This is at maximum capability and while the project is in its infancy, 400 turbines mounted on the seabed are predicted to generate 1.9GW. This is 16,000 GW hours per year, which would account for nearly half of Scotland's energy consumption.

How do they work? Just like wind turbines, but underwater. Because water is denser than air the blades for tidal turbines can be much smaller to generate the same amount of electricity – a good thing for the underwater wildlife passing through.

Maximum power west is 4 hours after high water Dover

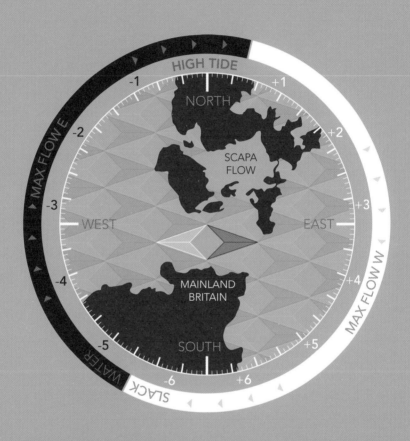

HIGH TIDE

NORTH

-1 +1

-2 +2

SCAPA
FLOW

MAX FLOW E

-3 +3

WEST EAST

MAX FLOW W

-4 +4

MAINLAND
BRITAIN

-5 +5

SLACK WATER

-6 +6

SOUTH

Maximum power east is 3 hours before high water Dover

Featuring a coastline of more than 17,700km and some of the strongest tidal streams in the world, Britain has a vast wealth of tidal power just waiting to be harnessed. While the sun doesn't always shine and the wind doesn't always blow, the tide always flows [except for slack water]. The best thing about tidal power is you can predict exactly how much you will be generating every day, for years in advance.

One way of converting the kinetic energy of tides into electricity is a barrage. This is best suited to estuaries such as the Humber, Mersey and Severn. The idea of a barrage is to work like a dam but with underwater turbines along its length. As the tide floods in and ebbs out the water is funnelled through the turbines, generating electricity. The Severn Barrage is the most developed plan in Britain with 216 turbines producing an average 2000MW – enough to power 6% of Britain. That's equal to 18 million tonnes of coal or 3 nuclear reactors.

A barrage may seem like an environmentally friendly form of generating energy, but think again. The Severn is one of Britain's leading bird habitats and the barrage would prevent the low-tide exposure of mudflats that sustain 85,000 migratory and winter wading birds. On top of this there are fears of severe silting and the worst damage of all [in the eyes of surfers] would be the eradication of one of the tidal wonders of the world – the Severn Bore. Many people feel a tidal stream power station should work harmoniously with nature in all aspects, so the engineers are going to have to find a less obtrusive form of harnessing the Severn. Across the estuary in Swansea they are attempting just this with a tidal lagoon that holds back the flood and ebb tides before releasing water when there is a height difference and the flow generates electricity for the grid.

There are many smaller scale options available and the most developed alternative is the horizontal axis turbine as used in Pentland Firth. There are also highly imaginative designs including huge corkscrews mounted onto the seabed that rotate as water flows through them, underwater kites that fly in a figure of eight with a turbine mounted below the wings and a hydrofoil that pulses up and down as water flows past. While these designs are full of inventiveness, they must be able to produce cost effective electricity for the masses if they are going to take off as a viable power solution.

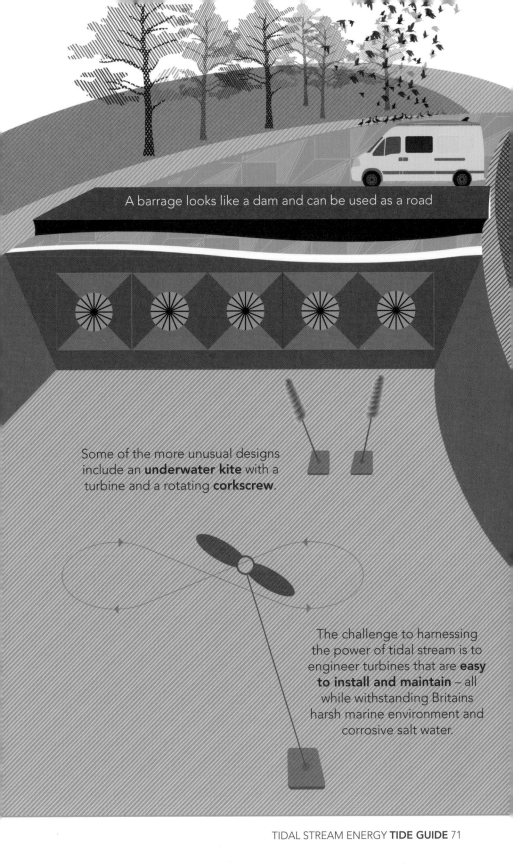

A barrage looks like a dam and can be used as a road

Some of the more unusual designs include an **underwater kite** with a turbine and a rotating **corkscrew**.

The challenge to harnessing the power of tidal stream is to engineer turbines that are **easy to install and maintain** – all while withstanding Britains harsh marine environment and corrosive salt water.

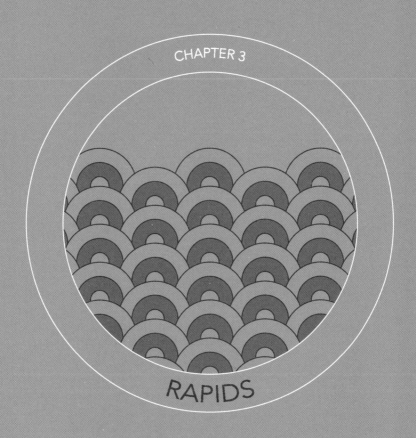

CHAPTER 3

RAPIDS

rapids *the turbulent motion of water*

When tide or stream are funnelled into a narrowing, the flow speeds up, and when this encounters a constriction tidal rapids are made. Put simply, place white water rapids into the sea and you have the tidal variety. Individual features include waves, eddies and in extreme cases – whirlpools [these have a section all to themselves, Chapter 4].

The constriction that makes rapids can either be the land, the seabed or both. Underwater obstacles can be shallow reefs rising up from the deep, while land constrictions can be channels between islands or headlands. Britain has many types of rapid and we're going to explore them in this chapter – from the Bitches in the south to the Merry Men of Mey in the north [both are far more fierce than their names suggest].

Rapids evolve hour by hour as the direction and strength of stream changes. On spring tides, the sea can transform from a calm haven at slack water to a bubbling, foaming and roaring cauldron at maximum flow just three hours later. Whether rapids are good or bad depends who you are and what you're in. Rapids have been known to capsize and sink large boats, and sailors should avoid them at all costs. If you're a swimmer without a flotation device you will probably drown in rapids. But if you're an experienced kayaker you'll have the time of your life surfing the standing waves that form behind rocky ledges. It's also possible to stand up and surf these waves, but it's much riskier if you fall in – which easily happens when there are so many hazardous currents around.

Whatever your view of rapids, it holds true that when wind and swell turn against the stream they become a very serious hazard with huge waves breaking from all directions and without warning. One of Britain's infamous rapids – the Portland Race – was once described as 'the master terror of our world'. This description could apply to any one of the rapids in this chapter if you catch them on the wrong day.

REEF
SANDBANK
ROCK SHELF

Spotting rapids is fairly easy – the challenge is to identify the individual elements within. Technically rapids are a collection of three of more whitewater features and these can be standing waves, holes, wave trains, eddies, eddylines and whirlpools. We're saving the next chapter for whirlpools, but here we're going to explore how to spot the other features and their causes and effects.

Standing waves are what the name suggests – waves that stand in one place. They are made when fast-flowing water runs down an underwater ramp. They are found at the bottom of the ramp when the water rises up into a clean glassy face with a patch of whitewater on the top. These are the *pièce-de-résistance* within tidal rapids and, while regular waves often only allow surfers rides of around ten seconds, standing waves can let you carve up and down the face for more than 10 minutes [or longer].

Holes are the naughty sibling of standing waves. While standing waves are 'unbroken' the whole top half of holes are 'breaking' and water is actually flowing back upstream. This is why they are also called stoppers – because when you paddle into them the water flowing upstream will literally stop you in your tracks. This aggressive current usually capsizes the unwary kayaker, which is why they are best avoided. Large holes are known as overfalls.

Wave trains are a collection of three or more standing waves or holes. They are usually irregular and bumpy, making them the whitewater equivalent to a rollercoaster.

Eddies are circular flowing calm areas downstream of obstructions such as rocks, pier legs or headlands. As the water flows against the obstruction it is compressed and flows past at high speeds. This leaves an area of low pressure behind the obstruction and the river fills this gap by pushing water back upstream. This is the purpose of the eddy. In French, they are called contre-courrant which is translated descriptively as 'counter current'. Eddies are safe havens within the turmoil of rapids, so a good place to rest and regroup.

Eddylines are a potentially dangerous line of water where the slow-moving eddy meets the fast flowing stream. The shape of the obstruction determines the shape of the eddyline. Sharp obstructions make harsh eddylines while soft obstructions make gentle eddylines. Where the eddyline is particularly concentrated the two opposing streams can create whirlpools.

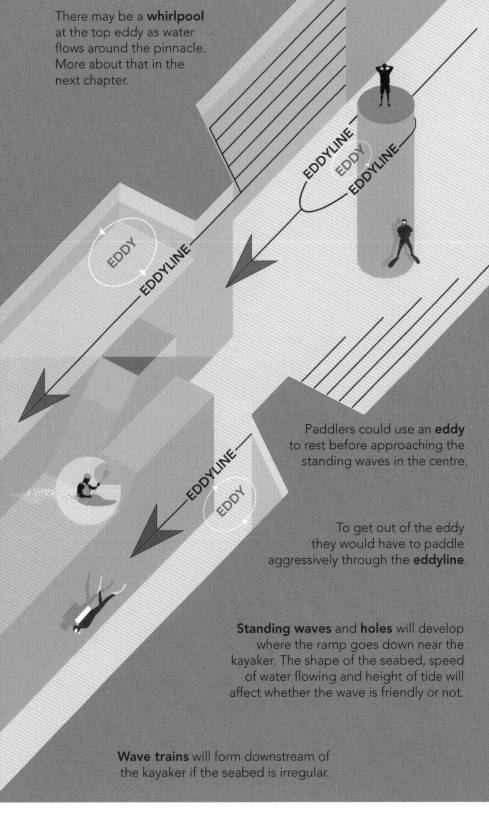

There may be a **whirlpool** at the top eddy as water flows around the pinnacle. More about that in the next chapter.

EDDYLINE

EDDY

EDDYLINE

EDDY

EDDYLINE

EDDYLINE

EDDY

Paddlers could use an **eddy** to rest before approaching the standing waves in the centre.

To get out of the eddy they would have to paddle aggressively through the **eddyline**.

Standing waves and **holes** will develop where the ramp goes down near the kayaker. The shape of the seabed, speed of water flowing and height of tide will affect whether the wave is friendly or not.

Wave trains will form downstream of the kayaker if the seabed is irregular.

The Bitches [I assume the name is dog related because the whelps are nearby] is a reef in Ramsey Sound just off the coast of St Davids in Pembrokeshire. As the tide rises in the Sound, the reef acts like a dam and because the stream flows north on the west coast at high tide, water is trying to get past the Bitches. When the tide becomes high enough water pours over the reef at speeds of 18 knots and when this comes into contact with the underwater obstacles standing waves and eddies are formed. This makes them a popular spot for playboaters – kayakers who like to surf the glassy waves here.

The conditions are not always picture perfect. In the early hours of 13 October 1910 the St Davids Lifeboat – the Gem – was called out to rescue a crew of three from the ketch 'Democrat' which was in danger of being swept onto the reef. Strong winds and swell created a terrible sea and it took the lifeboat an hour of paddling to reach the Democrat. In challenging conditions, they successfully picked up the crew but on the way back wind and stream overcame their tired arms and they couldn't paddle fast enough to avoid the Bitches. The Gem was forced onto the reef and three of the lifeboat crew died. The 15 survivors clung onto the higher rocks for 12 hours while the rapids roared below them. When the rapids ebbed down with the tide the survivors were rescued eventually by two local boats.

The lifeboat disaster highlights the dangers of the Bitches and they haven't calmed down much in the last 100 years. Playboaters knocked out of their kayaks have been sucked under the eddy lines only to reappear half a minute later and 50 metres downstream. These are only minor hazards compared to Horse Rock which threatens tired paddlers on their way back from a play session. The granite pinnacle rises vertically up to the surface from 60 metres below, and furious whirlpools develop when the stream flows past. These are not an abstract danger but a very real threat, as one unlucky playboater discovered with tragic consequences. [Read more about this on page 108.]

THE BITCHES

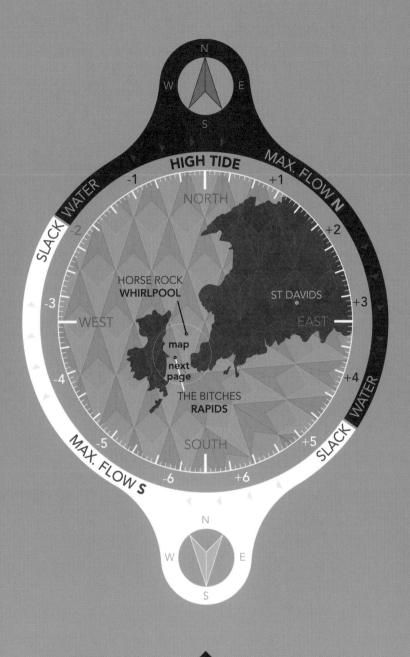

The rapids are best on the flood tide when water flows north

The Bitches reef creates such sublime standing waves that playboaters from all around the world test their skill here. In 1991, the first ever Kayak Rodeo World Championships were held on these rapids. It's a challenging environment and definitely not for beginners – even experienced paddlers should go in groups so they can keep an eye out for each other. Although mainly enjoyed by kayakers, surfers are starting to share the waves too.

High tides of 6 metres makes the standing waves but the higher the better – **7 metre tides** on Springs make special waves! Low air pressure will make the tide and waves even higher.

Top Wave forms just past the highest rock and although it's narrow, the clean glassy wave is popular with playboaters.

Bitches Hole is a stopper found close to Ramsey Island. It's a friendy wave in calm conditions but can turn into a terror when the swell is against the stream.

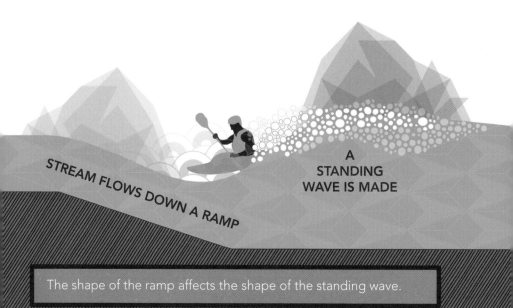

STREAM FLOWS DOWN A RAMP

A STANDING WAVE IS MADE

The shape of the ramp affects the shape of the standing wave.

The paddle out to the bitches takes 25 minutes. Plan to enter the water just over **3 hours before high tide** Milford Haven in order to arrive when they are in full flow.

Watch out for deadly **whirlpools** at horse rock on the return journey.

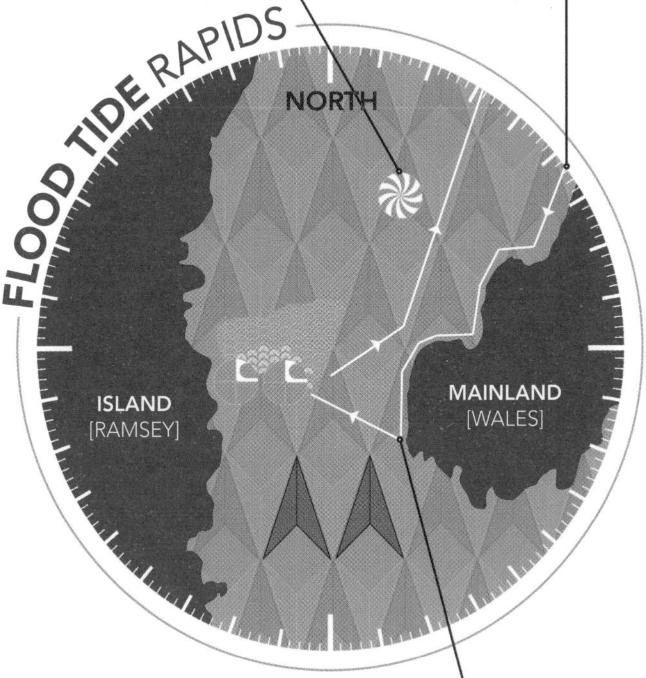

FLOOD TIDE RAPIDS

NORTH

ISLAND
[RAMSEY]

MAINLAND
[WALES]

When you can't get any further upstream, **ferry glide** [a kayaking technique of crossing fast flowing rivers] across the sound. When you get to the other side an eddy will catch you and take you upstream where you're ready to take on the Bitches.

DEEP WATER AROUND THE RAPIDS IS CALM

Portland Race was once described as 'the master terror of our world'. Before you ask why a Race is in the Rapids chapter, it's because race is another word for rapids. And Portland is a particularly violent tidal rapid that should be avoided at all costs. Despite the danger there is always someone prepared to take the risk and a small contingency of scuba divers frequent these waters when they're in the mood for being thrown around in a huge washing machine.

The race starts off the tip of Portland Bill in Dorset and can extend around 3km south. Unlike other rapids that only happen when the stream is flowing one way, Portland works on both streams which means there's double the trouble. But what creates the rapids? At around half an hour before high tide in Weymouth the water in Lyme Bay begins to flow east. When the water closer to shore comes into contact with Portland it is deflected south and builds speed. The stream further out to sea has avoided the obstacle so is still flowing due east, but when the two streams converge the sea becomes turbulent. The west-going stream works in the same principle but from the opposite direction.

Almost a kilometre off the bill is the Portland Ledge where the seabed rises up from 30 metres to 10 metres. This dramatic obstruction forces the 10-knot stream up to the surface. The water breaks in heavy waves made even more erratic by the irregular seabed of the ledge. This is only the beginning – when wind or swell oppose the direction of stream the Portland Race becomes 'the master terror' that is capable of sinking even big boats.

For this reason passing sailors are advised to venture no closer than 4.8km from the Bill in fine weather and 12.8km in bad weather. For those wanting a quick way around Portland there is a kilometre wide channel between the bill and ledge where the sea is calmer due to the deeper water. However, there's no room for error because if the wind drops or your motor stops, the south flowing stream will drag you into the race. If so you've just got to hope your boat is shipshape and strong enough for the abuse.

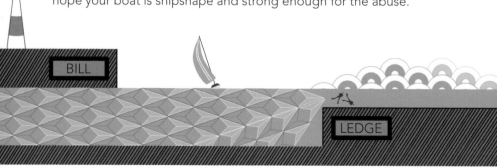

BILL

LEDGE

The race happens on both **east and west** streams

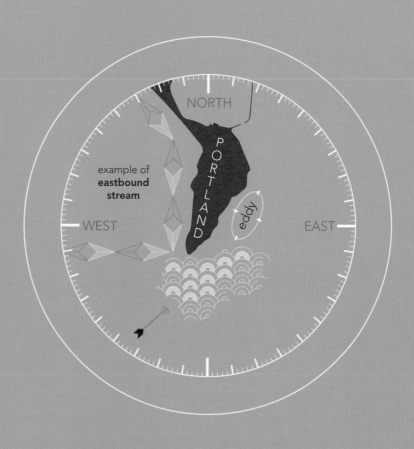

A large eddy forms on the opposite side to the stream direction

The Falls of Lora are playful tidal rapids where Loch Etive meets the sea near Oban in north-west Scotland. Don't mistake the word playful – they can be deadly – but while other rapids around Britain invoke pure terror, these ones provide a playground for experienced kayakers and scuba divers.

The Falls are created by a narrowing and shallowing at the entrance to Loch Etive. As the tide drops at sea, 66-million cubic litres of water must drain from the loch through the constriction, creating fast flowing streams of up to 12 knots. When these come into contact with the shallow seabed standing waves and eddy's are formed. The effect is tumultuous for both spectators on the shore and adventurers in the water, made even more dramatic by the 1.2-metre height difference that can be seen on either side of the rapids as the water level at sea drops faster than in the loch. The rapids also work on the flood tide, but while the principle is the same they are less powerful.

The height of the tide is crucial to how powerful the rapids are. On the ebb-tide rapids, lower tides are better. This is because the tidal difference on either side becomes greater, so the water flows through with more pressure. On the flood-tide rapids, higher tides are better because the water level at sea rises higher than in the loch. This means the rapids are best on spring tides when the tidal range is greatest. Spring tides are indeed crucial, but wind and air pressure are also important.

High air pressure and offshore winds make lower tides so are the best conditions for the ebb-tide rapids. Low air pressure and onshore winds are better for the flood-tide rapids because they create higher tides. Although spring tides can be predicted years in advance the wind and air pressure can only be forecast a few days before. This adds to the unpredictability – and anticipation – of how high the Falls are going to be flowing.

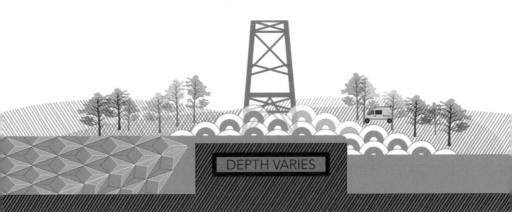

DEPTH VARIES

The flood-tide rapids are strongest ½ hour before high tide

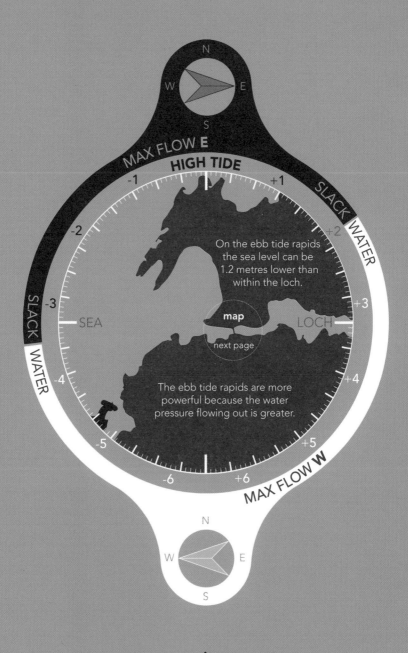

On the ebb tide rapids the sea level can be 1.2 metres lower than within the loch.

map
next page

The ebb tide rapids are more powerful because the water pressure flowing out is greater.

The ebb-tide rapids are strongest 5 ½ hours after high tide

The Falls of Lora are popular for adventures both on and below the surface. For whitewater kayakers the conditions are best on the ebb as this is when the Falls are most powerful. For seakayakers the flood tide is preferable because the waves are shallower and less aggressive. For scuba divers, slack water on the flood tide is the best time but you're in for a rough ride even then.

Flood tide is best for kayak training sessions as you don't need to worry about anyone being swept out to sea.

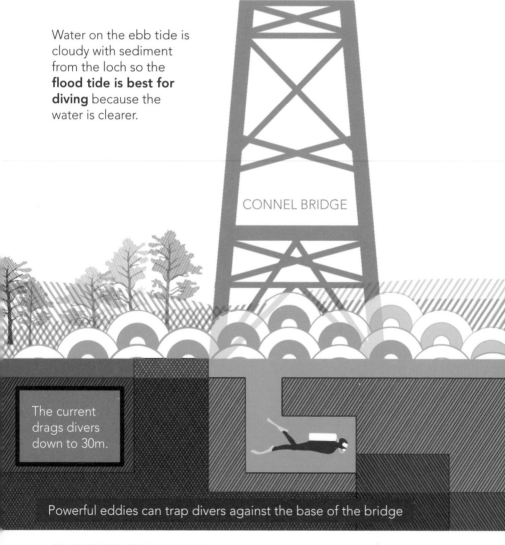

Water on the ebb tide is cloudy with sediment from the loch so the **flood tide is best for diving** because the water is clearer.

CONNEL BRIDGE

The current drags divers down to 30m.

Powerful eddies can trap divers against the base of the bridge

There are three waves to kayak surf – the **main wave** which starts nice but grows into a stopper, the **centre wave** and the clean green **forever wave**. Take your pick!

Kayakers must watch out for **whirlpools** on the eddy line just down from the waves.

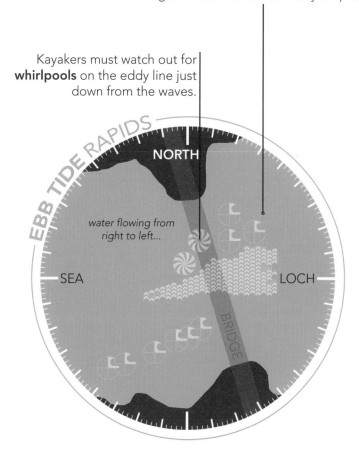

EBB TIDE RAPIDS

NORTH

water flowing from right to left...

SEA

LOCH

BRIDGE

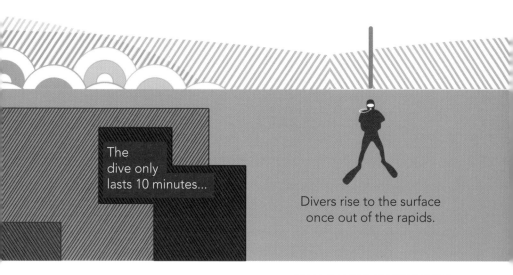

The dive only lasts 10 minutes...

Divers rise to the surface once out of the rapids.

With some of the most powerful tidal streams in the world and a scattering of islands and underwater obstructions, it can come as no surprise that the Pentland Firth is tidal rapid heaven [or more appropriately hell]. Of the many rapids that form here the most dangerous is the Merry Men of Mey – and there's nothing merry about them.

The race starts small and grows as the west flowing stream speeds up. It first forms off the Men of Mey rocks half-an-hour after high water Dover [tidal information is often linked here to avoid the complications of local tides for sailors] and at full stream extends all the way across the Firth. Within this huge area the most dangerous spot is above a sand wave field [a sandy seabed that has the shape of waves] 5½ kilometres west of Stroma where large wave trains form over the undulating seabed. The waves are biggest on spring tides when the flow is fastest but when swell pulses in from the Atlantic the opposing bodies of water create an unimaginably terrifying sea. At these times monster waves emerge out of nowhere and crash down from both directions, making it a miserable experience for anyone unfortunate enough to find themselves here at the wrong time and tide.

While the Merry Men may seem like an indefatigable beast, remember they have a weakness – stream. Without stream flowing through the firth there is no energy to power the race. This may come as no comfort to someone found in there at maximum flow: 'just hold on until slack water in three hours'. But it is not just slack water that tames the beast – the Merry Men don't even form on the east-flowing flood tide. The only problem is that when they are resting the other races in the Pentland Firth are just waking up.

Some of the biggest dangers in the firth are eddies. While the eddy's themselves are friendly, the eddylines that form where the fast flowing stream meets the slow moving eddy can prove fatal. If a vessel finds itself caught in an eddyline, the current will spin the vessel around. This could result in capsizing and sinking. With all these hazards it is no wonder sailors are recommended to wait until neap tides – once a fortnight – until crossing.

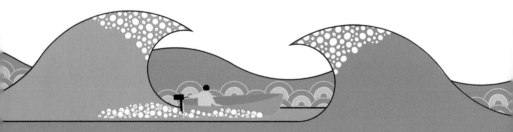

The race begins **half an hour after** high water Dover

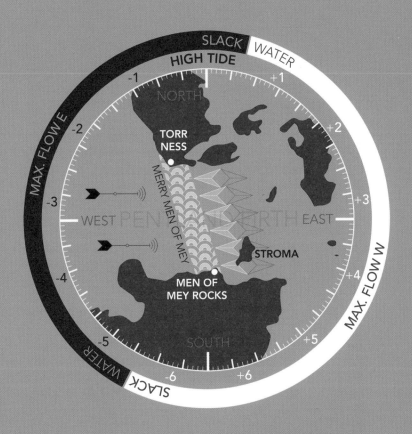

The race calms down around **5½ hours before** high water Dover

The Swellies is the narrowest part of the Menai Straits [between Anglesey and mainland Wales] and is generally regarded as the waters between the Menai and Britannia suspension bridges. On spring tides streams of eight knots flow past islets and over rocky ledges, forming tidal rapid features. By now you should be able to imagine what is coming – eddies downstream of the islets and waves over the rocky ledges.

For the boats passing through the Menai Straits, the obstacles and their rapids are a hazard best avoided. But for kayakers this is what they have travelled all this way for. With both types of boat looking for different conditions the function of this 'sea river' changes with the power of flow.

When the flow is weakest at slack water the sailors have a window to pass through. But they can only use the slack nearer high tide [1½ hours before] because at slack nearer low tide there isn't enough water in the straits and rocks lie perilously close to the surface – half a metre in some places. This gives boats just one brief window to pass through every 12 hours, 25 minutes. If they miss this it can be too dangerous to pass because the flow is speeding up and rapids are forming – much to the joy of the kayakers waiting on the banks.

As the flow speeds up the waves grow bigger and the eddylines become more pronounced. For inexperienced playboaters this can come as a shock, but for the regulars they can punch through the turbulent water to the shelter of the eddy. But it is the standing waves they have come for, and the best ones are found at Gored Goch and Swellies Rock. It would be wrong to say these were consistent because, as we have discovered with other tidal rapids around Britain, air pressure and wind can have a big impact on the conditions of the day [much to the frustration of the paddlers who have travelled a long way to get here]. But if they have a bad session it's not the end of the world because on spring tides the rapids can be surfed at breakfast, lunch and tea so there's always the chance to have another go later.

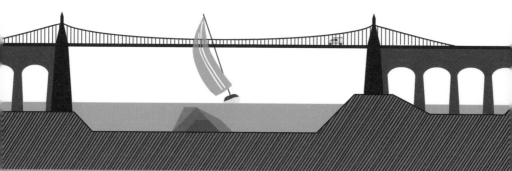

Sailing boats pass through at slack 1½ hours before high

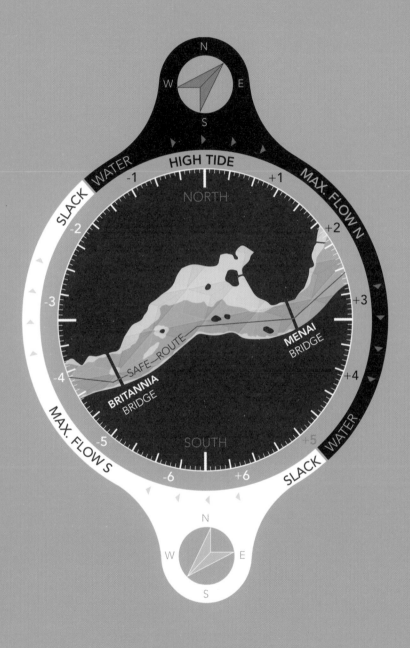

Kayakers are waiting for maximum flow 2 hours after high

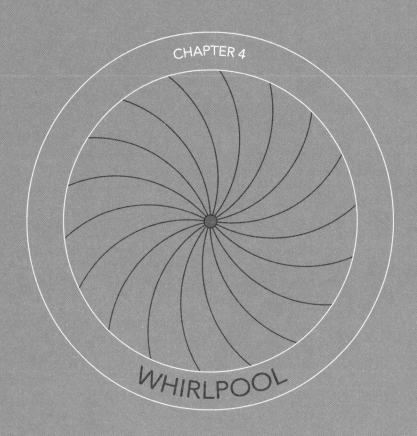

CHAPTER 4

WHIRLPOOL

whirlpool the downward vortex of water

Of all the flows of water around Britain, whirlpools surely have the strongest pull on our primeval fear of the sea. Can there be a worse moment than being swept powerlessly towards the vortex where the only route is down, down, down? This is what happened to a mannequin that was dropped into the Corryvreckan [the third largest whirlpool in the world] as an experiment. The dummy was recovered downstream and the depth gauge attached to the lifejacket showed it had been dragged along the seabed 84 metres below the surface.

If whirlpools are so dangerous why do we not hear more stories of people being sucked under? The answer is simple – because they exude such visual and audible terror that nobody in their right mind would go near one at full power. That excludes George Orwell [the author of *Animal Farm* and *1984*] who found himself caught up in a maelstrom so powerful it sheared off the outboard motor to his dinghy (see p.100–1). Orwell was not the first to encounter the terror of British whirlpools. They also had a profound impact on the Vikings who plundered this island in the first-century AD. Although the Scandinavians were highly advanced mariners, the physics of whirlpools baffled them. To make sense of these waters they developed stories of mythical witches, cauldrons, heroes and sea monsters.

Myths aside, what actually makes whirlpools? They happen when contrasting currents meet and wrap around each other to create a circular-flowing body of water. If there is enough intensity a vortex will form and the whirlpool will develop its downward pull. Most whirlpools at sea happen in areas where tidal rapids are found and fast flowing streams encounter an unusually shaped seabed. But they also happen when another terror of the sea – the tsunami – surges over land and in this rare event monster whirlpools have been recorded around the world. Could this happen in Britain? Let's find out.

Listen. You are more likely to hear a whirlpool before you see it – especially if it is one of the big ones. Across the Atlantic in Canada the Old Sow was named because of the strange sucking noises it makes. In Scotland, the Corryvreckan simply roars. If a swell from the Atlantic is pulsing in at the time of maximum flow, and during spring tides, the maelstrom can be heard from many miles away. This sound is the interaction between the water molecules as the crash, collide and vortex.

Look out for circular motions of water. When you get close enough to see the whirlpool on a calm day, the swirling circular flows of water will be clearly visible. In places the centre of the whirlpool is noticeably lower and spirals down into the depths. But this is very rare because when a wind or swell converges with the fast flowing streams, standing waves and turbulent waters are created. At these times the whirlpool merges with the features of tidal rapids. For this reason, rapids are a good place to look for whirlpools.

Look out for eddylines. Because whirlpools form when opposing currents meet, eddylines can be home to small whirlpools where the fast flowing stream interacts with the slow moving counter-clockwise eddy. As eddys form downstream of obstacles, these are features to look out for. Islands or large rocks can provide suitable conditions but the challenge comes when they are submerged. But even if you can't see the obstacle, the eddylines and wavetrains will be clearly visible on the surface. Use your knowledge of rapids to imagine what is beneath the surface.

Look at maximum flow on spring tides. The danger is that you look at slack water. Because there is no stream there is no eddy, no eddyline and no whirlpool. But if you were to return just three hours later the conditions would be appreciably different. For this reason it would be sensible to return every few hours – or even every few days – to notice the difference between slow and fast streams, as well as spring and neap tides.

The Scottish have a myth that the hag goddess of winter – Cailleach – washes her plaid [blanket] in the Gulf of Corryvreckan at the beginning of winter. This three-day event can be heard from over 30 kilometres away and when she has finished, the cloth is pure white and becomes the snow that covers the land. There is some insight here. Usually the Corryvreckan whirlpool can he heard from ten miles away but at times such as the winter equinox, the position of the sun results in especially powerful spring tides. These turn the biggest whirlpool in Britain [and third largest in the world] into overdrive, with the roaring water clearly audible from twenty miles.

But what makes this mythical water feature? Most of the magic happens below the surface. Here the flood tide water is funnelled through the Sound of Jura and it flows through a deep trench, down a 219-metre hole, then up a steep pinnacle that rises to within 29 metres of the surface. When this powerful up-thrust of water meets the west flowing surface stream the opposing currents create whirlpools and vortices. In case you didn't know, vortices are whirlpools with a downward motion and they're the ones we're all terrified about.

On a windless day the opposing currents and subsequent vortices can be clearly seen whirling below the glassy surface. However, when wind or swell comes in from the west the opposing bodies of energy create standing waves that can grow up to 10 metres. The glassy surface is transformed into a roaring cauldron.

Back in the day of Norsemen, a young prince called Breackan anchored his boat in the whirlpool for three days and nights to prove his love for the local chieftain's daughter. With his vessel secured to the shore by three ropes – one made from hemp, another from wool and the third from maidens' hair, he braved the maelstrom. On the first day the first rope snapped, on the second day the wool rope gave way but the maidens' hair remained strong until the third night when it snapped under the strain and Breackan was drowned in the whirlpool. Racked by guilt, one of the maidens admitted she was not as pure as originally claimed.

The biggest tides don't always produce the biggest whirlpools

▼

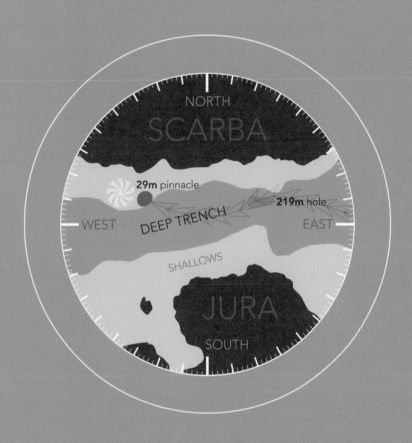

NORTH

SCARBA

29m pinnacle

219m hole

WEST

DEEP TRENCH

EAST

SHALLOWS

JURA

SOUTH

▲

Wind, air pressure and swell affect the conditions on the day

If the heroic Breackan could not survive Corryvreckan, would you believe me if I told you that George Orwell, weakened by tuberculosis, tumbled through the maelstrom in a 14"-dinghy crewed by his young nephew, niece and 3-year-old son?

You may well wonder what the famous author was doing there in the first place. In the summer of 1947 Orwell was staying in a farmhouse on Jura to focus on his novel, *1984*. For a break, he invited his nephew and nieces to visit, treating them to a boating expedition. It is unknown whether he checked the tides at all, or simply misjudged them, but when the small dinghy came around the headland it was swept into the maelstrom. With a strong swell pulsing in from Atlantic the Gulf was in full swing with all the frills – standing waves, eddy's, eddylines, lots of small whirlpools and some big ones for good measure. In the turbulence their outboard engine was shaken off and dropped into the vortex. Staying quite calm, Orwell sat in the back and instructed his nephew, 'Motors gone. Better get the oars out, Hen. Can't help much I'm afraid.'

Luckily the stream flowing through the Sound of Jura was slowing down and the whirlpool was receding. This allowed the motley crew to escape to the deserted island of Eilean Mor, a kilometre from Jura. Here, the same swell that frenzied the whirlpool was pitching the boat up and down by 4 metres, causing the dinghy to capsize near the shore. Orwell was trapped beneath the upturned boat but managed to swim out with his 3-year-old son in his arms.

With a battered dinghy, one oar, a fishing rod and a wet cigarette lighter, Orwell seemed to thrive on the desert island experience and went in search of food [although they had eaten breakfast an hour before] while the others waved the fishing rod with a shirt on it. After an hour and a half a passing fishing boat spotted the distress signal and picked them up. A lucky escape! Or was Big Brother keeping the author alive just long enough to complete his book?

The Corryvreckan misadventure and subsequent immersion in cold water is thought to have enhanced the tuberculosis and Orwell's health deteriorated over the next two years as he feverishly worked on completing his masterpiece. He died months after its publication – just long enough to appreciate the overwhelming response.

The Corryvreckan has been described as **Britain's most dangerous dive** that should only be attempted when you have the experience of 500 dives under your belt.

Divers drop 25kg of shot down onto the pinnacle connected to a rope that is tied to 2 x 25-litre buoys. They use the line to get down to the pinnacle.

Time on the pinnacle can be as little as **5 minutes** before the vortex picks up again.

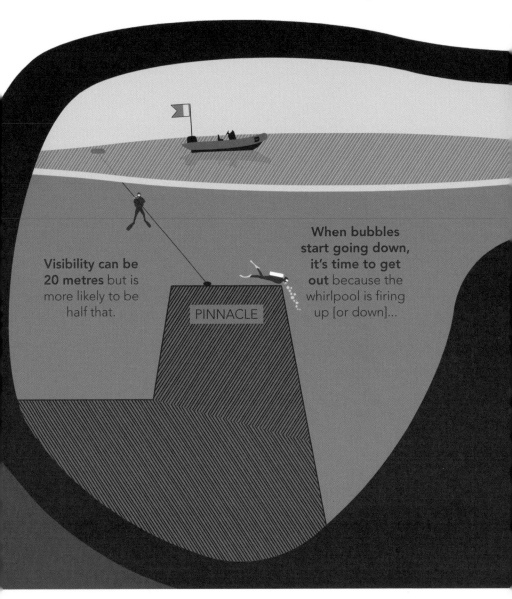

Visibility can be **20 metres** but is more likely to be half that.

PINNACLE

When bubbles start going down, it's time to get out because the whirlpool is firing up [or down]...

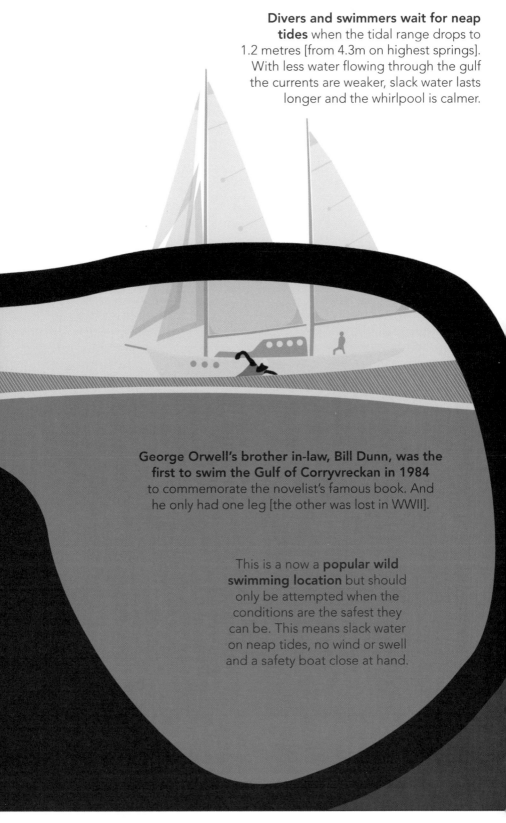

Divers and swimmers wait for neap tides when the tidal range drops to 1.2 metres [from 4.3m on highest springs]. With less water flowing through the gulf the currents are weaker, slack water lasts longer and the whirlpool is calmer.

George Orwell's brother in-law, Bill Dunn, was the first to swim the Gulf of Corryvreckan in 1984 to commemorate the novelist's famous book. And he only had one leg [the other was lost in WWII].

This is a now a **popular wild swimming location** but should only be attempted when the conditions are the safest they can be. This means slack water on neap tides, no wind or swell and a safety boat close at hand.

The west coast of Scotland is one of my favourite places in Britain because, in a rare display of chivalry, humans have taken a back-seat and let nature run the show. Or it could be that few of us are hardy enough to live up there. Either way, nature reigns in this land and Corryvreckan is one of the kings. His dominion is clearly defined by the steep walls of the gulf and his throne rests upon the pinnacle where the whirlpool forms. Of his loyal subjects you can find an entire ecosystem including foraging predators such as the minke whale, porpoise, gannet and gull.

The shags and guillemots know this already, but the west and east flowing streams create very different conditions. The violent whirlpools and standing waves mainly happen during the west stream, and this drives mixed waters deep down into the Firth of Lorne. In contrast, the east flowing stream brings nutrient rich waters and prey up to the surface where the predators sweep in for supper. Don't be mistaken into thinking this is a feeding frenzy – far from it. This ecosystem enjoys a clearly defined structure where different predators ebb and flow with the tide. With individual skill sets, hunting techniques and dietary tastes it makes sense that the variety of seabirds and cetaceans [whales and dolphins] prefer different conditions.

The large gulls are the least tactical and can be found in all places at any time. The shags can also be found at all tides but they focus on the slow moving eddys that form on both east and west streams. Because their swimming speed of 2 metres per second is half the maximum flow of the main channel they sensibly avoid the fast currents. Although porpoises can handle the faster water they prefer the slower streams and are rarely seen on spring tides. This is more likely to be because their food is not around then. The birds that do like the fast flowing currents are also the most abundant in this kingdom – auks and kittiwakes. The polar opposite are the black guillemots who, like us humans, only venture out at slack water.

We still don't know what animals brave the whirlpool at maximum vortex as, unsurprisingly, nobody has ever gone down there at full flow to see.

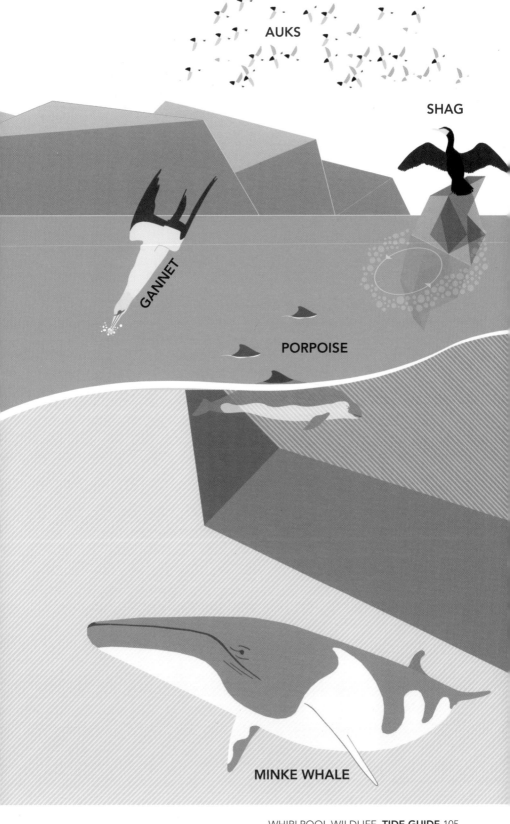

AUKS

SHAG

GANNET

PORPOISE

MINKE WHALE

Stroma is a wild and windswept island in the middle of the Pentland Firth and on its northern tip lies the Swilkie whirlpool. The name Stroma derives from the Norse word 'Straumr' meaning 'island in the stream' and it is this stream flowing against the island that makes the whirlpool. Unlike other whirlpools around Britain, the Swilkie is equally powerful regardless of which direction the stream in flowing, although it does calm down a bit at slack water.

As the fast flowing waters of the Pentland Firth hit Stroma they are compressed and flow around the northern tip at high speeds. This leaves an area of low pressure on the other side of the island, and the gap is filled by water flowing 'upstream'. This is an eddy. Where the calm eddy meets the fast flowing stream on the northern tip of the island the contrasting currents create powerful eddylines and the Swilkie whirlpool is found in these.

Swilkie is derived from another Norse word – 'Svalgr' – which somewhat terrifyingly means 'the swallower'. The Vikings blamed the Svalgr for many shipwrecks and developed a myth that both explained this phenomena and warned sailors to stay well away. The story tells how King Frodi enslaved two giantesses Fenia and Menia to grind whatever he demanded on a magical quern called Grotti. The giantesses were not at all happy about their forced labour so used the quern to grind an army to free them from Frodi. With the help of the sea king Mysing they sailed away, but disaster struck in the Pentland Firth when Mysing asked Fenia and Menia to grind salt. This they did, but in such vast quantities that their longship sank under the weight of it.

Grotti still lies on the seabed off the northern tip of Stroma and it continues to grind the salt that makes the sea saline. Sailors must be careful not to venture too close to Stroma because the eye of the whirlpool is directly above the eye of the quern, where the seawater is sucked down to grind into salt.

The whirlpool is calmer **half an hour after** high water Dover

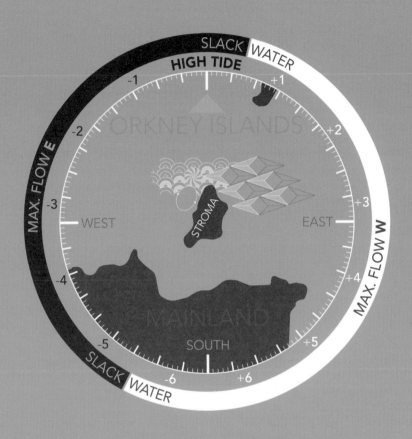

The quern takes a short break **5½ hours before** high water Dover

In the previous chapter (page 78) we discovered how playboaters [kayakers] flock to the Bitches in Wales to test their skill on the rapids that form when a fast flowing stream pours over the reef. Just downstream of the playspots is another quirk of nature, but instead of excitement this one sends shivers down the backs of anyone in the water. Welcome to Horse Rock and its whirlpools.

The hazard is marked as 0.9 metres above the lowest spring tides, which means most of the time it is underwater. Despite this it can clearly be located by the wavetrains and boiling eddylines that form above it. This turbulent water is formed by the spire shape of Horse Rock, rising up from 60 metres below the surface. This is a very similar bathymetry [seabed shape] to the Corryvreckan and the whirlpools are likely formed in the same fashion, with the opposing currents created by the presence of the rock creating powerful eddys and turbulent eddylines.

A hazard best avoided, you may think. The only problem is that kayakers must pass 80 metres upstream of the rock on their route back from the Bitches. It's a nerve-wracking paddle and made even worse by the knowledge that a kayaker was once taken under and didn't come back up alive. There is an alternative route that passes downstream of the rock but there's a risk of being caught in the back eddy and swept back in towards the whirlpools. Most opt to pass upstream and paddle for their lives.

Horse Rock is not just a hazard for small vessels. In September 2000, a 10-metre rib [rigid hull inflatable boat] was having fun in Ramsey Sound when they passed too close to the rock. The boat found itself surfing a wave towards the turbulent waters and when the bow [front] of the boat hit the eddyline it was sharply swept to the port [left] by the opposing current. This irregular movement caused the boat to pitch over and capsize. All fourteen occupants were thrown into the water. Remarkably they managed to get back to the upturned boat and hold on before anyone was sucked under. Another rib in the area spotted the disaster and quickly picked them up.

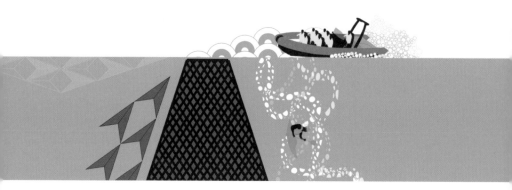

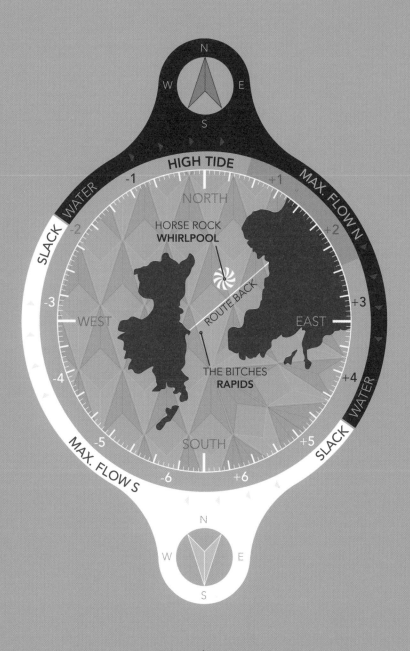

The north stream pushes playboaters towards the whirlpool

If reading about whirlpools made you nervous you will be relieved to hear the chapter is almost over. Now we can get to the fun – waves and bores. But before that we must squeeze in another terror of the sea. Tsunami. And to finish off this chapter, let's blend the two together and explore whirlpools generated by tsunamis.

Before you say tsunamis don't happen in Britain, read the next chapter then come back to this page. Because tsunamis do happen. But luckily they are very rare – especially the big ones. While whirlpools generated by tsunamis have never been seen on our shores, there is a chance it may happen in the future. But what makes them? When a tsunami surges inland it floods huge areas of dry land and structures that are usually far away from the waters edge temporarily become the seabed. When the tsunami waters recede the huge volumes of turbulent water flow aggressively against these obstacles and create opposing currents. As we have learnt in this chapter, two bodies of water flowing in different directions will wrap around each other and create a whirlpool. If the power of the water is strong enough a vortex will form.

In 1755, a powerful earthquake struck off the coast of Portugal and the tsunami waves it generated hit Cornwall. But they were most intense in Portugal's capital, Lisbon, where huge whirlpools formed in the harbour. Onlookers record seeing boats swept into the vortex and not coming back up again. That would make it more powerful than the Corryvreckan. Similar whirlpools were recorded during the tsunamis in the Indian Ocean in 2005 and Japan in 2011.

The good news is that these whirlpools don't last long because they recede with the water from the tsunami. When they are gone they will never return again, becoming surreal memories that can scarcely be imagined. But the big question is whether a tsunami large enough to power this monster maelstrom is possible in Britain. Let's find out.

WATER DRAINS OUT

As water drains out to sea, counter currents generate whirlpools

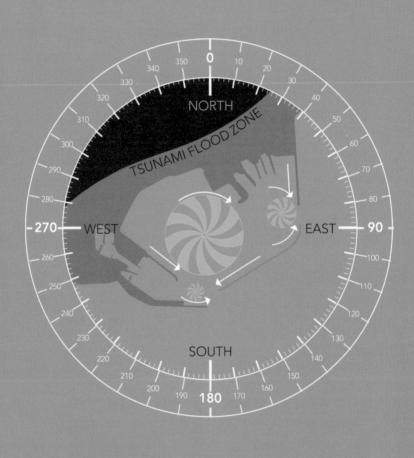

Please take note – this is only a theoretical phenomenon

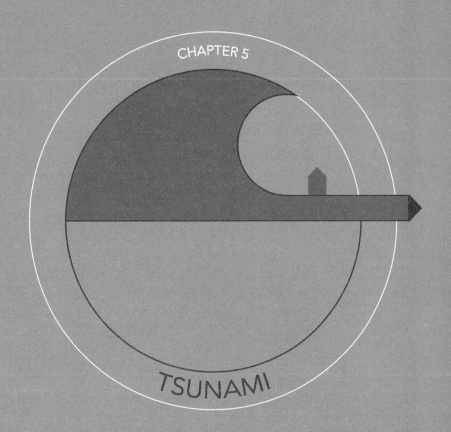

CHAPTER 5

TSUNAMI

tsunami *a surge of water inland*

Tsunami in Britain? Surely not! But if you think about it, why is this any more far fetched than Scotland covered in ice or men running around Kent in loincloths? Yes, both did happen. And it would have been one of these mesolothic hunters who experienced Britain's biggest tsunami 8,000 years ago. Since then there have been a handful more but the most important to you and I are the ones scientists think may happen in the future.

There are many myths about tsunamis and even their alternative name 'tidal wave' is misleading. They actually have nothing to do with tides, although a clear sign that one is approaching is a dramatic fall in the sea level in just a few minutes – even lower than the lowest spring tide. This is the trough of the wave and the peak usually follows around 5 minutes later. So resist your inquisitiveness and get to safety [see page 126, How to survive a tsunami].

The statistics are both awe-inspiring and terrifying. They can travel 800kmh over deep water. That's the speed of a jet plane. And when they approach shallow water the bottom half slows down and the top half pitches up to heights reaching 30 metres [a ten-storey building]. Unlike regular waves a tsunami doesn't stop on the beach but surges onwards through the lowest lying land, taking cars, trees and buildings with it. At this point the real danger is being caught in the churning water.

Many people think tsunamis are a single wave. This is not true – sometimes three waves can happen over a period of 24 hours and the first is not always the largest. They also have the ability to 'wrap' around headlands so not only exposed coastlines are at risk. But that's enough of the scary bit because it's not the purpose of this chapter. Instead, I hope the following pages will give you a greater awareness of the signs and an understanding of what to do in the very, very rare possibility that you find a tsunami [the Japanese word for 'harbour' and 'wave'] bearing down on your beach.

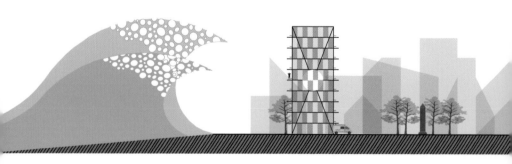

Around the world there are universal natural and man-made warnings that a tsunami is approaching. Because these waves are so rare in Britain, the man-made systems are less developed than in other parts of the world. This becomes apparent if you travel to the Pacific Ring of Fire [where 80% of tsunamis happen]. There you can download an app, subscribe to email alerts, listen out for sirens located in low lying areas at risk and then follow the escape route signs to safety. Back in Britain, you will have to be alert for nature's signs and work out your own safety plan.

Animals will be acting strangely. If you are scuba diving you will notice the marine life behaving differently to their usual patterns – they may even disappear altogether. Animals on land also pick up the signals quicker than humans and dogs often act strangely [whining, barking or hiding] minutes before a tsunami. Another clear sign is birds flying away from low land.

If you feel, hear [of] or see an earthquake, meteorite, landslide or volcanic eruption a tsunami may soon follow. Luckily Britain doesn't have any active volcanoes and earthquakes are generally too small to generate tsunami. However if these happen far away in the Atlantic Ocean or North Sea the waves can still affect Britain. Remember that a tsunami will travel 2,000 miles across an ocean in just four hours.

The clearest indicator down on the beach is a fast receding tide. In just a few minutes the sea level can drop lower than the lowest spring tides and fish caught out by the strong current will be lying on what is usually the seabed. Because the tide falls lower than usual, there may even be never seen shipwrecks exposed. This suction of water is the trough of the wave and the peak follows a few minutes later.

Because the waves can wrap around headlands you may hear a tsunami before you see it. The sound is similar to a freight train. When it arrives people often find it's not what they expected because it is not always the classic wave shape. Instead, a tsunami can look like a fast rising tide that flows up the beach and over harbour walls.

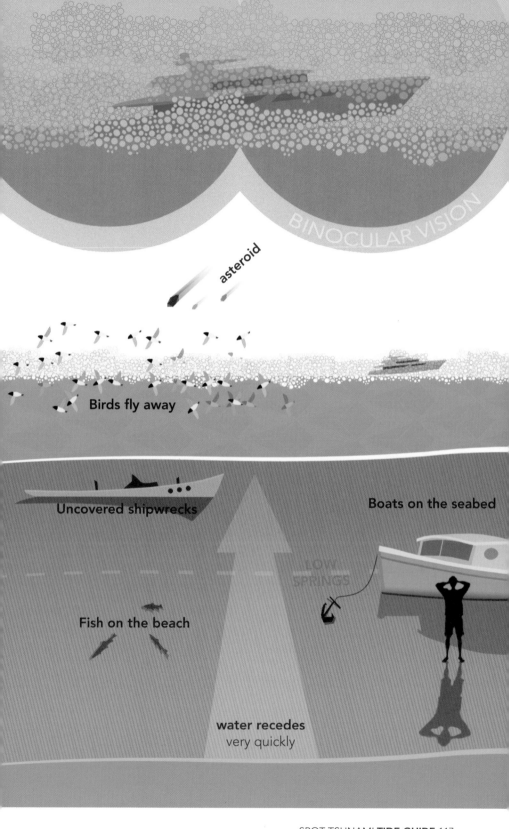

asteroid

BINOCULAR VISION

Birds fly away

Uncovered shipwrecks

Boats on the seabed

LOW SPRINGS

Fish on the beach

water recedes
very quickly

There are four main natural causes of tsunami – earthquake, landslide, volcanic activity or a meteorite striking the ocean surface. Earthquakes account for the majority of these waves.

Earthquake. Contrary to popular belief, earthquakes do happen in Britain and the British Geological Survey measures over a hundred tremors every year. Of these, around 25 are felt by people. Britain's biggest ever quake was near Dogger Bank in 1931 and measured 6.1 on the Richter Scale. Scientists believe 6.5 is the greatest magnitude earthquake possible in Britain, while 7.0 is widely believed to be the minimum required to set off a tsunami. The reason for our reduced earthquake activity is because they are most frequent at subduction zones where tectonic plates meet. The closest danger zone is the Azores–Gibraltar fault line. This runs between Africa and Europe and joins the mid-Atlantic ridge [900 miles away]. A tsunami triggered from an earthquake there will take about 5 hours to reach Britain.

Landslide. Both submarine [under water] and terrestrial [above water] landslides can cause tsunamis. These can be created by several factors including earthquakes or severe weathering of the slope during storms. The biggest tsunami in human occupation of Britain was caused by a submarine landslide off the coast of Norway that set off a 20-metre wave bearing down the North Sea 8,000 years ago.

Volcano. There are several volcanoes in Britain but they are all extinct – the youngest is 60 million years old. This is because we are no longer in an area of tectonic activity. But we are in threat of volcanic eruptions from places such as the Canary Islands where an eruption could send out pyroclastic flows of hot rock, pumice, ash and gas. This, or the collapse of an erupting or inactive volcano, would set off a tsunami through the Atlantic.

Meteorite. This is the rarest cause, but is believed to be responsible for a massive tsunami that resulted in the extinction of dinosaurs. According to NASA, once every 2,000 years a meteoroid the size of a cricket pitch passes through the atmosphere and if it lands in the ocean [four-fifths of the earth's surface] this will cause a tsunami. Britain would be most at threat from an impact in the North Sea or Atlantic, but with modern technology we should be able to deflect it off course.

Earthquakes are the most frequent cause of tsunami

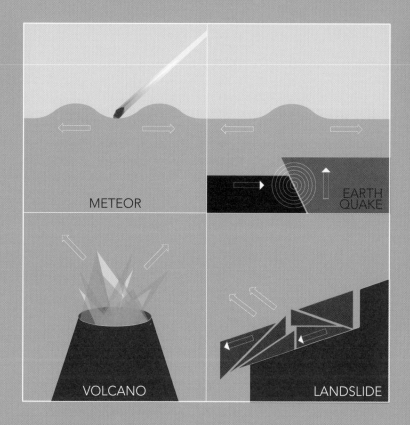

METEOR

EARTH QUAKE

VOLCANO

LANDSLIDE

A **landslide** off Norway, 8,000 years ago, made Britain's biggest tsunami

Britain has not always been an island.

Ten thousand years ago there was a land bridge connecting the east coast with Denmark and the Netherlands, allowing the free movement of people and goods back and forth between Britain and Europe. But rising sea levels caused by melting ice caps [it was the end of the last ice age] were gradually flooding this utopian hunting ground known as Doggerland.

The final blow which severed the physical connection between Britain and Europe was a tsunami that swept over Doggerland 8,200 years ago. This catastrophe was the result of a submarine landslide off Norway that is believed to have been triggered by either an earthquake or a sudden release of methane gas trapped during the last ice age. Either way, the 'Storegga Slide' displaced 3,000 cubic kilometres of sediment and the tsunami this set in motion changed Britain's identity forever.

The waves that hit the Shetland Islands reached 20 metres in height. Due to the enormous power that created them, they continued down the east coast and tsunamite [sediment from a tsunami] can be found 40km inland and 6 metres above the present sea level. That would have been 20 metres above the high tide line at the time.

While the first wave was tearing down the east coast, the inhabitants of Doggerland would have been preparing for winter after having spent the summer hunting elk in the mountains of Europe. It is unknown whether they would have recognised nature's signs of the wave approaching but, even if they had, they could not escape to safety because the highest point of Doggerland was only 5 metres above the sea.

In a cruel blow of nature, Doggerland and its inhabitants were drowned by the tsunami and now their story lies at the bottom of the North Sea. But through the destruction of this land the island of Britain became fully formed and our identity as a seafaring nation began. Could this be British history's defining moment?

The Storegga Slide caused Britain's biggest tsunami in **6,200BC**

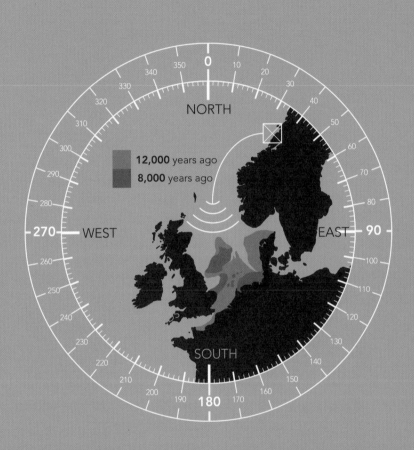

The wave was **20-metres high** when it hit the Shetland Islands

Tsunamis are not always the monster waves we are led to believe. Sometimes they look like a fast rising tide that surges inland and this flow of water is very similar to a more frequent event in Britain – the storm surge. These happen when a high spring tide coincides with low air pressure and strong onshore winds [see page 30, How wind & air pressure affect tide] and the effect can be disastrous. The biggest storm surge happened in 1953 and was described as 'the worst peacetime disaster to hit Britain'.

The east coast is most susceptible to storm surge. This is because typically deep depressions pass north of Scotland from the Atlantic and the low pressure results in an increase in the height of the North Sea – 1cm per 1 milibar drop in pressure. The storm also drives Northerly winds, with the resulting 'wind drift' forcing water south towards the English Channel. But when it reaches the Strait of Dover the flow is restricted by the narrow [35km wide] gap and this leads to a build up in the southern North Sea. The water level can rise by over 3 metres and when this coincides with high spring tides the sea becomes higher than the land. Added to this, the strong northerly winds build waves that pulse even higher than the sea level, so increase the flooding.

This exact process happened in 1953 and, again, sixty years later in 2013. While both surges caused widespread flooding, the earlier event resulted in the much higher death toll of 325 in Britain. This was due to a combination of extremely low pressure [964hPa], powerful winds gusting 126kph, 5.6-metre higher tides than usual, and the ineffective governmental system of warning and evacuating communities at risk. Even if the local population had been adequately prepared, the destruction would still have been widespread with damage to power stations, transport links, water systems, sewage pipes, and the flooding of 160,000 acres. In total, the cost was calculated at £50 million. In today's money that's £1.2 billion.

By 2013, lessons had been learnt. The sea defences had been strengthened and early warning systems developed. The government knew what was coming and an effective evacuation plan was in place. There were far less casualties this time around but widespread damage still resulted with lots of land beneath houses and beach huts being washed away by the sea.

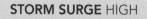

STORM SURGE HIGH

NORMAL TIDE

Low air pressure system off Scotland increases North Sea tides

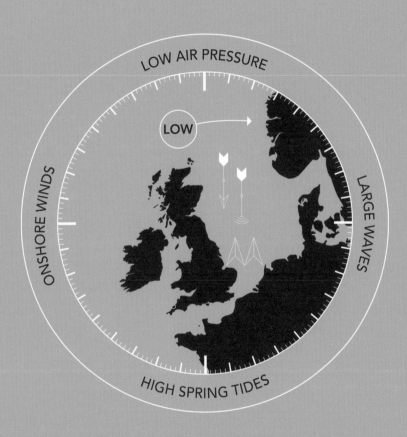

LOW AIR PRESSURE

LOW

ONSHORE WINDS

LARGE WAVES

HIGH SPRING TIDES

Northerly winds push water down onto low lying area of Norfolk

At 9.40 on the morning of 1 November 1755 a massive earth-quake measuring 8.5 on the Richter scale struck 193km south-west of Lisbon [the capital of Portugal]. This caused widespread damage throughout the crowded city and locals flocked down to the open harbour to escape the fires and falling rubble. Many even piled onto boats. A sensible idea? As it turned out, no. Forty minutes later, a tsunami generated by the earthquake tore through the harbour, laying waste to the boats and waterfront.

If this is a book about the British coast, why am I writing about a tsunami that happened over 1,600 kilometres away in Portugal, you may ask? Well, apart from the warning never to go down to the sea after an earthquake, this tsunami did not stop in Lisbon. Instead, it pulsed out across the Atlantic and hit the coasts of the Caribbean, Brazil, North Africa – and Cornwall [five hours after the earthquake occurred].

While the biggest waves of 20 metres were recorded in Morocco, Cornwall was battered by a series of 3–4 metre waves for about two hours that afternoon. The waves were largest in Mounts Bay because of the gradual shallowing of the seabed while a 3-metre surge in the tide was recorded in Newlyn. St Ives and Hayle also experienced this surge and the effects of the tsunami were felt in south Wales and as far away as the Thames. While there are many reports of the strange behaviour from the sea, there is no documented survey of the damage. The closest description we get is 'a great loss of life and property occurred upon the coasts of Cornwall' which is frustratingly vague. Perhaps this is because Britain had bigger concerns at the time – the Seven Year's War was just starting.

Because the earthquake happened on an established fault line it is likely another may strike again, but it is unlikely to be anywhere near as large as in 1755. While the media have sensationalised that a tsunami could 'sweep the Scilly Isles off the map' scientists believe the waves would be 2 metres, so similar to the scale of storm surge we are now becoming used to in Britain.

The tsunami was triggered by a magnitude 8.5 **earthquake**

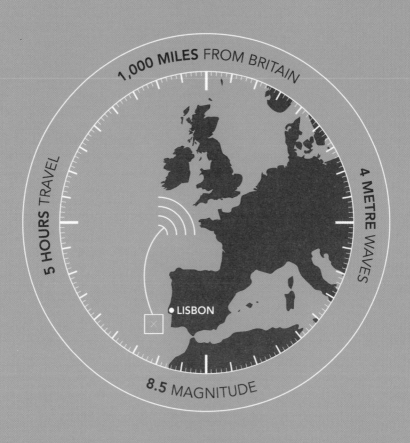

1,000 MILES FROM BRITAIN

5 HOURS TRAVEL

4 METRE WAVES

LISBON

8.5 MAGNITUDE

It took **5 hours** to reach Britain **1,609 kilometres** away

In the days after 27 June 2011 the British media reported a tsunami along the south coast of England, apparently caused by a submarine landslide on the continental shelf 200 miles south west of Cornwall. However, when the British Geological Survey studied the journey of the waves they discounted this theory. It must have been triggered by an earthquake then. But no earthquakes were recorded. No meteorites had hit the ocean. No volcanoes had erupted or collapsed. What could it have been then?

Dog walkers at St Michaels Mount in Cornwall recounted how the tidal causeway suddenly flooded and their hair was left standing on end. This is common with atmospheric disturbance and suggests a meteorological event – a meteotsunami. These look and move just like regular tsunami waves but are created by a rapid change in air pressure such as squall lines [a line of lighting and thunder that forms on the edges of cold fronts]. The sea level rises in the form of a wave beneath this weather front and the wave is driven along below the storm. So if storm surges are also created by low air pressure, what is the difference between the two? During a storm surge the drop in air pressure also raises the sea level, but it only floods coastal areas when strong onshore winds coincide with high spring tides.

So, this was definitely a meteotsunami and weather reports support the theory. The general area of low pressure originated in Spain and Portugal, before moving north-east through the English Channel. On closer inspection there were actually three weather fronts that created three independent meteotsunamis from the French–Spanish border in the Bay of Biscay, all the way up to the Strait of Dover. This sounds very dramatic but the waves were small and created little damage. It was more of a fascinating phenomenon and a video in the Yealm Estuary in Devon shows a 0.5 metre wave flowing upstream just like a tidal bore [see next chapter]. But with more severe storms as a result of climate change Britain could see more powerful meteotsunamis in the future. All it takes is a sudden drop of 3 milibar in the air pressure to create a 3cm wave at sea. When this is funnelled into a narrowing and shallowing coastline the waves have the capacity to grow up to 4 metres, with the potential to create much more damage than in 2011.

TIDAL CAUSEWAY

A **3cm** wave at sea can grow to **4m** in shallow estuaries

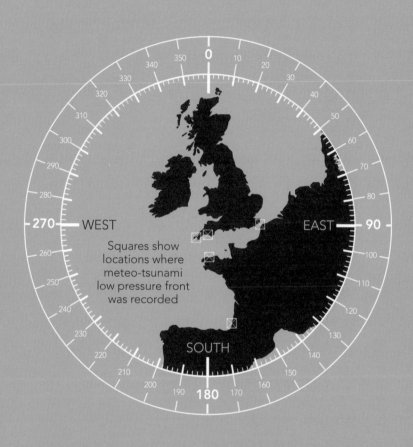

Squares show locations where meteo-tsunami low pressure front was recorded

WEST

EAST

SOUTH

Meteotsunamis are made by a **rapid change in air pressure**

Tsunamis are so rare in Britain that you will probably never need to know how to survive one, but in the off chance it does happen this knowledge will come in handy. And if you like to travel to exotic places such as the Pacific [where 80% of tsunamis occur] this information could save your life.

The secret to surviving a tsunami is to be quick. Spot the signs and get to safety [30 metres high or about 3 kilometres inland] avoiding river valleys. The key to speed is to be prepared – learn the natural warnings of an approaching tsunami [see page 114] and pre-plan your quickest route to safety. If there is a big earthquake [unlikely in Britain] the road may be damaged so it's a good idea to be aware of a route by foot as well as car. If you want to be especially prepared a rescue bag will pay dividends. In it you could have water, snacks, warm clothes, waterproof matches, a head torch and a solar charger for your phone.

If you are caught out unprepared and your first sign of the tsunami is hearing or seeing it, the chances are you won't have time to get to safe ground. Instead, get to the top of the biggest and sturdiest looking building and stay away from the side where the water is coming from. If that's not possible the next best thing is to climb a tree although there's a 50/50 chance it may snap. But the good news is the first wave is rarely the biggest so if the water recedes and you have the chance to get into a safer position quickly – take it. And if you were able to get to safe ground in the first place – stay there.

The worst-case scenario is that you find yourself in the water. If you are in a boat near deep water when the tsunami is approaching it may be safer to head out into even deeper water rather than risk going into the shallows where the waves will be larger. If you find yourself on shore and swept up by the turbulent water, try to get onto something big and floaty. This will keep your legs away from any debris, stop you becoming exhausted, keep you safe from small whirlpools and make you more noticeable to rescue services, if you get swept out to sea by the receding water.

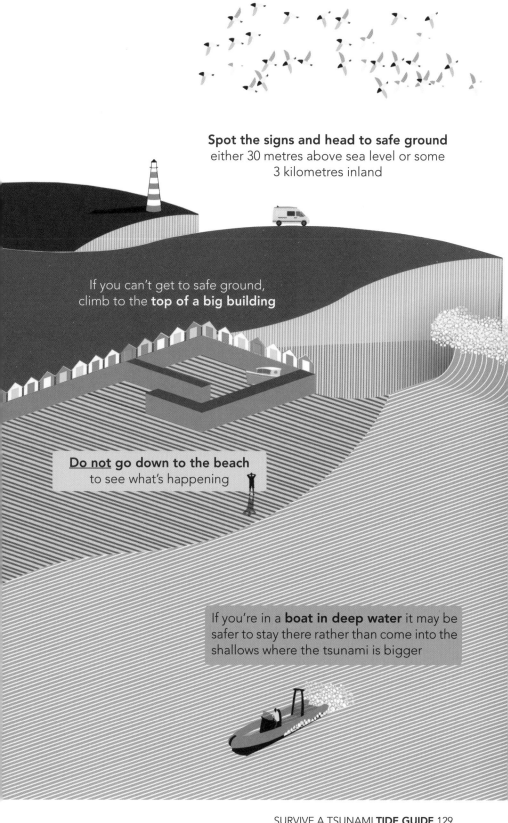

Spot the signs and head to safe ground
either 30 metres above sea level or some
3 kilometres inland

If you can't get to safe ground,
climb to the **top of a big building**

Do not go down to the beach
to see what's happening

If you're in a **boat in deep water** it may be
safer to stay there rather than come into the
shallows where the tsunami is bigger

On 30 June 2015 a research paper warned that Britain was at risk of an asteroid-triggered tsunami. The study coincided with the first ever Asteroid Day [set up to raise awareness of the threat from asteroids] and stated there was a 1 in 10,000 chance an asteroid would crash into the sea off Norfolk in the next 85 years. This is a terrifying thought. But as a percentage it becomes 0.01% which sounds much lower. And in those 85 years there are likely to be considerably more storm surges and meteotsunamis to deal with. However, a bigger asteroid hitting elsewhere in the world could damage Britain too. Three billion years ago a giant asteroid triggered a tsunami that travelled around the world *three times.*

Massive asteroids are very rare and it is more likely a 50-metre diameter rock would crash into the sea and trigger a 3-metre wave which would threaten low lying areas around Britain. The problem is that we are only aware of 1% of these asteroids and the founders of Asteroid Day have set up a 100x Declaration that is encouraging world bodies to step up research. The declaration calls to increase discovery by 100 times to 100,000 per year and this would mean in ten years time we could know about every tsunami-potential asteroid [there are 1 million of them]. Once a danger rock has been located scientists can then study the journey patterns and calculate an impact corridor and time it could strike earth – decades in advance. Ten years' warning should be enough to develop a plan to deflect the asteroid.

There are three main ways to deflect an asteroid away from earth. The gentlest technique is to position a large spaceship alongside the rock. Over time the gravitational pull from the ship on the rock should push the asteroid off course. If that doesn't work a spaceship could actually nudge the rock to change its destination. And in the most extreme situation we can always 'nuke' it. This is more delicate than it sounds because simply breaking a big rock into lots of little rocks would create new problems. It is much better to tactically place the nuclear bomb so that it maintains the structural integrity of the rock whilst pushing its away from earth. For all of these strategies the key is early warning – the sooner the asteroid can be located the gentler a nudge it needs.

The tsunami risk is from **North Sea** or **Atlantic** impact

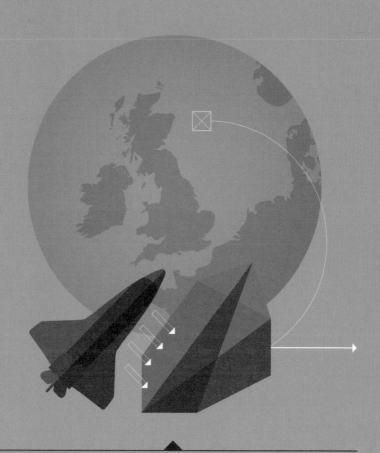

A **50-metre** diameter asteroid can make a **3-metre** high wave

There is a small island within the Canary Islands called La Palma, and within La Palma is the volcanic ridge of Cumbre Vieja. This tiny dot on the world map came to global attention when it was announced the volcano would soon erupt, collapse and generate a mega-tsunami. According to the scientists, there is a crack 25km long x 2km deep x 15km wide and when the volcano next erupts [anytime within the next few decades] the 500-cubic kilometre rock within this crack will fall into the Atlantic Ocean in just 90 seconds. This sudden displacement of water would create a 900 metre high wave. By the time the tsunami reaches Africa the waves will have reduced to 100 metres. On the eastern seaboard of the US the waves will measure 50 metres. And four hours after the rock fall, the first 40-metre wave will hit Britain and flow about 6 kilometres inland.

That's what the scientists said, but after this tsunami-scare more detailed studies were carried out by separate groups of scientists and they all came to the conclusion this hypothesis is impossible. Above land, they found the crack to be only 2.5km long and restricted to the surface. It would take 10,000 years for this shallow slice to develop into anything capable of creating such a large landslide. Other scientists got underwater and studied previous submarine landslides from the Cumbre Vieja. They all showed that the past events had occurred in small sections and over a long period of time. This is more likely to happen in the future and each small landslide is potentially only capable of a localised tsunami – if that.

Despite the overwhelming criticism to the 'mega-tsunami' theory, it is difficult to shake off such catastrophic imagery. Just like the whirlpool, tsunamis appear in our consciousness as an unstoppable force of nature that we are simply powerless against. Despite the statistics that 'you are more likely to be killed by a coconut landing on your head', our imagination keeps returning to the tsunami. It must be an intrinsic part of human nature and although it may never happen to me, I'll be packing a rescue bag just in case. At the very least, it makes me feel better.

A minority of scientists warn of a mega-tsunami

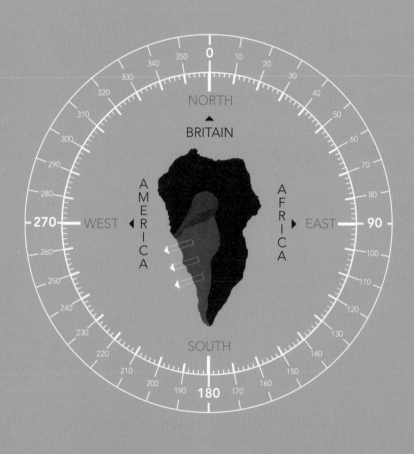

A majority of scientists disagree with the mega-tsunami theory

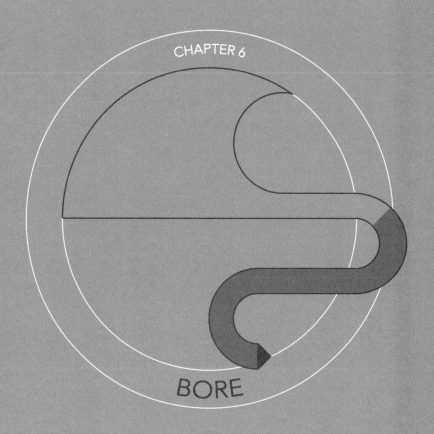

CHAPTER 6

BORE

bore *a wave flowing up a river*

I am sorry if the last two chapters made for stressful reading. The good news is we can now get back to the fun and what better way than to enjoy a true natural wonder – the tidal bore. If you were surprised to hear that water flows up rivers towards high tide [Page 26, Tidal Thames] you'll be astonished to discover waves as high as 2 metres pulse many kilometres up some British waterways. These bores are very rare [less than 100 are known around the world] and one-fifth of them are found in Britain. That's because we tick all the right boxes.

There are two criteria that make tidal bores. Firstly, the tidal range should generally be greater than 6 metres. Secondly, there must be a wide and flat estuary or bay that funnels water into a narrow and shallow river. The shape of the bay intensifies the height of the tide and holds back the flood until it 'bursts' forward with greater power. When this surge is tapered, a wave is formed. And at this moment, oh joy of joys – it can be surfed! While beach waves only offer surfers rides measured in tens of metres, bores can be surfed for kilometres. The Severn bore is Britain's longest wave and the record distance surfed is over 14.5km.

Some bores happen every day, but most form during spring tides [after full or new moons] when the higher tidal range causes water to 'spring' forth with more energy. Autumn and spring equinoxes are even better. Although tides are the most important ingredient of a 'good bore' – air pressure, wind and rainfall all impact the speed and power of the wave. But bores are not just a single wave. Following close behind is a huge body of water – the rising tide – and many secondary waves, known as whelps, are found here.

Bores are not just magnets for humans wanting to savour nature. They also provide rich feeding and breeding grounds for marine life. The problem for fish is they often find themselves dazed and confused when the bores tumble through and this is when the predators – sharks, piranhas, crocodiles – sweep in. Luckily we don't have too many of these in Britain so the biggest danger for surfers is colliding into each other on the most crowded 'party waves' in Britain.

At the very beginning of this book, I mentioned how the vertical motion of tide is so subtle that it is impossible to see with the naked eye. While I enjoyed a relaxing afternoon in Lyme Regis attempting to disprove this theory, I must admit it is possibly the least exciting hobby anyone could take up. But if you hold back that tide, then release it like a slingshot, the surge of water in the form of a bore becomes a dramatic spectacle enjoyed by thousands of people around Britain – and the world – every fortnight.

If you would like to get into bore watching, you're on the right island. Britain must have the greatest concentration of bores in the world and they have been found on the Rivers Severn, Trent, Kent, Dee, Mersey, Parrett, Welland, Great Ouse, Ouse, Eden, Esk, Nith, Lune, Ribble and Yealm. While each bore has its own unique characteristics [height, speed, time and tide] there are some universal rules that create a better bore. Higher tides are generally a key component and these can be found around spring tides and equinoxes. But sometimes even on days with lower tides, there can be bigger bores. This will be because of onshore winds and low air pressure on the day, as well as low rainfall over the preceding ones. So check your tide table first, then look at the barometer, wind vane and rainfall gauge.

Choosing a spot to watch the bore takes some thought and the ultimate in planning would be to follow it as it travels up the river. This can be achieved by choosing your first spot nearer the estuary [downstream] that is easily accessible by road. When the bore passes, you can then jump in the car, overtake the wave and intercept it further upstream. Bridges are great for this.

Once you have judged the weather conditions to be perfect and planned your viewing points, the final challenge is to arrive early. Not only does this ensure you don't miss the bore – it allows you to soak in the atmosphere of the slow moving river. This will make it even more shocking when the bore arrives, providing a moment you will never forget.

Tidal bores generally happen just **before high spring tides**

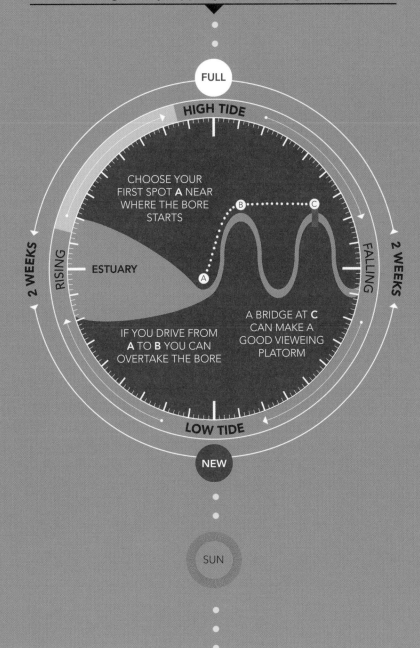

FULL

HIGH TIDE

CHOOSE YOUR
FIRST SPOT **A** NEAR
WHERE THE BORE
STARTS

B ··········· C

ESTUARY

A

2 WEEKS

RISING

FALLING

2 WEEKS

A BRIDGE AT **C**
CAN MAKE A
GOOD VIEWEING
PLATORM

IF YOU DRIVE FROM
A TO **B** YOU CAN
OVERTAKE THE BORE

LOW TIDE

NEW

SUN

The River Severn produces Britain's biggest bore. For surfers this makes it *the best*. Its potential for surfing was first exploited in 1955 when Colonel Jack Churchill, an ex-commando with a bravery medal [from WWII: not for surfing the bore] paddled out into the river on a board he designed especially for the occasion. By catching his first river wave, 'Mad Jack' became the first person in the world to ever surf a tidal bore. Since then, watermen and women from all around Britain have flocked to the Severn to ride the longest waves of their lives. The record lasted over an hour [imagine the thigh burn].

You don't need to be a surfer to enjoy this wonder of nature. For many, the bore is invigorating enough just to watch. But don't get too close to the riverbank because the breaking waves could drag you in [see p.144 – Bore dangers]. The good news is you're unlikely to get caught off guard because you will hear the bore long before you see it [see page 150 – Hearing bores].

The Severn Bore happens about 260 times a year. That's twice a day for 130 days. Of these, 25 days are most suitable for surfing when the tide is higher than 9.5 metres at Sharpness. But it is not just tide that affects the size of the bore. If it has been raining heavily the river is deeper and the waves don't grow as tall. Wind and air pressure also affect the wave dramatically. Low air pressure and south-westerly winds will allow the bore to travel faster upriver and it may arrive half an hour before anticipated. Contrastingly, high air pressure and north-easterly winds can hold back the waves for up to 30 minutes. This means the bore can arrive anytime within an hour. And when it does, if the conditions are just right, the wave can reach peaks of over 2 metres and will flow all the way up to Maisemore [beyond Gloucester] where the weir finally brings the bore to a halt.

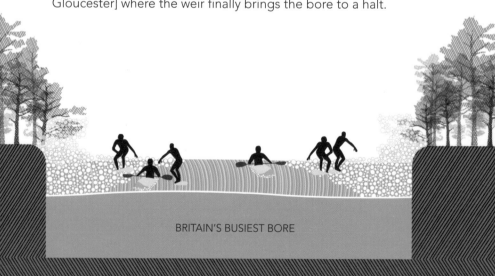

BRITAIN'S BUSIEST BORE

The Severn is home to **Britain's biggest – and busiest – bore**

NORTH

GLOUCESTER

BORE STARTS

a 15-metre tide is funnelled
up the Severn Estuary

WEST

270

CARDIFF

EAST

90

BRISTOL

SEVERN ESTUARY

SOUTH

The front wave is followed by whelps and the rising tide

The Severn Bore is biggest [best] on spring tides when the tide is higher than 9.5 metres in Sharpness. The ideal conditions are low rainfall, low air pressure and a south-westerly wind. This will form waves around 2 metres tall, while the biggest waves recorded were 2.8 metres on 15 October 1966.

WARNING THIS ADVENTURE IS NOT FOR BEGINNERS
[read about the dangers of bores on the next page]

The bore happens **day and night** [between 7 and 12] and although it is not recommended to surf in the dark, some people love it.

The record distance surfed is over 14.4km. While this is a challenge of strength and balance, you must also judge the wave and follow the crest that moves across the river.

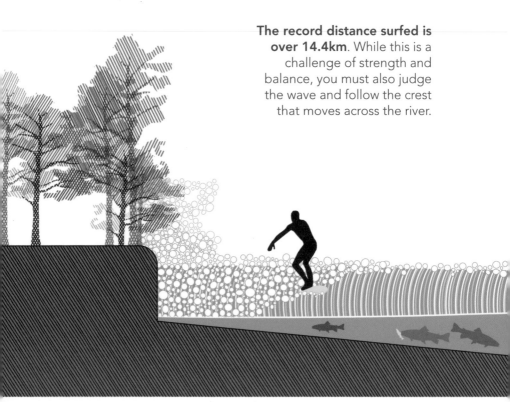

To catch and surf the wave, you'll need a **longboard** which is why stand-up paddleboards are perfect for the challenge. You can also use your paddle to measure the depth of water [the wave breaks in shallower water].

Boots will protect your feet not only from the cold but also from sharp objects in the river and banks.

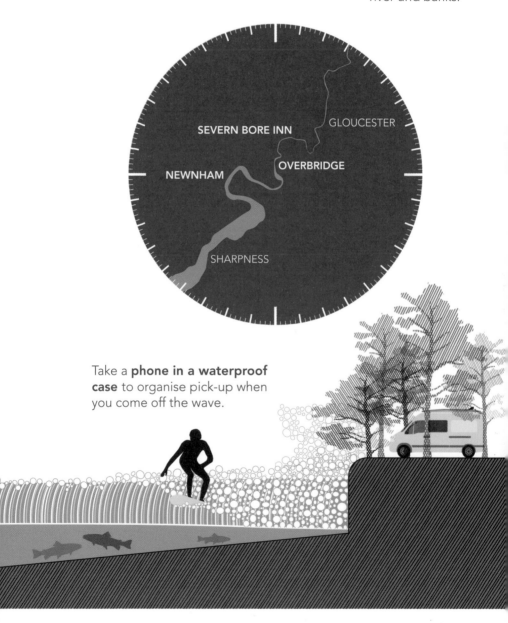

Take a **phone in a waterproof case** to organise pick-up when you come off the wave.

Bores are not just dangerous for surfers; they can be deadly for spectators too. Every year a handful of people around the world drown when tidal bores surge up the riverbank, knock them off their feet and sweep them into the turbulent water. Although British bores may be smaller than others around the world, it doesn't take long for hypothermia to kick in on a cold January night and there are numerous hazards to avoid on the river bed, in the water and on the riverbank. So take a step back from the edge.

Dangers in the water. The obvious danger is sustaining a head injury from being hit by a surfboard or a kayak. The Severn gets so crowded it's a surprise this doesn't happen more often. The secondary dangers are the submerged hazards in the 'main flow' of the bore behind the wave. Trees are often torn from the riverbanks and there have also been sightings of fridges, gas canisters, wooden pallets and car tyres following the wave upstream. And if you avoid all of these there is a risk of being run over by one of the many ribs [speedboats] that follow the wave. So make yourself visible when you're in the water.

Dangers on the riverbed. Many 'industrial' rivers around Britain will show the signs of human occupation. The biggest dangers may be the jagged remains of shipwrecks or steel bars protruding from reinforced concrete. There will also be the natural obstacles such as rocky outcrops lying just beneath the surface in parts of the rivers.

Dangers on the riverbank. The main dangers are low trees and their branches. When the wave surges up onto the banks it can push surfers towards the undergrowth where there is a danger of being knocked out by a limb of a tree or getting your leash tangled in a branch and becoming trapped. There may also be hazards such as quicksand or crumbling cliffs when getting to and from the river. These are easier to avoid when getting in because you can plan where to enter the water. But when you come off the wave you will hopefully be miles up river and you may be somewhere the riverbanks make safe exit very difficult, especially if you are contending with a strong currents.

The secret to safe and successful bore surfing? Precise planning and preparation.

Wearing boots will protect your feet from sharp objects on the riverbanks.

Make yourself visible to boats following the wave upstream

Don't get too close to the riverbanks as you could crash into a tree or get your leash caught on a branch.

Research the underwater dangers before you surf the bore. There may be the remains of shipwrecks, steel bars protruding from concrete or rocky outcrops lying just beneath the surface.

While the Severn is at the forefront of modern British bore surfing, the Trent Aegir is steeped in ancient mystery. Why is it called an 'Aegir', what does the word mean? And is it true that the Danish King Canute attempted [unsuccessfully] to hold back the tide on this very bore a thousand years ago?

Despite the strange name its physical characteristics are the same as other bores around Britain. As the peak of the tide wave flows south down the east coast, water is held back by the Humber Estuary until it bursts upriver. Here, the shallowing and narrowing of the River Trent forces the front of the bore up into a wave. This surge of water travels 80 kilometres to Britain's most inland port, Gainsborough. It used to travel further until dredging and weirs blocked its progress.

There are many theories as to what 'Aegir' means or comes from. My favourite is from Norse mythology. Aegir and his wife Ran, were the gods of the sea. They had nine daughters, including Bara (the Norse word from which 'bore' originates). So, in colloquial English, Aegir is the daddy of all bores. With peaks averaging 1.5 metres it is smaller than the Severn, but perhaps the Vikings named it on a day with particularly low air pressure, low rainfall and strong easterly winds. These would all conspire to create a bigger and more powerful wave that would have travelled even further inland, especially without the modern man-made obstructions in the river.

If the name 'Aegir' is open to gentle interpretation, the claim that King Canute played his tide game here is a point of hot debate. This is because there are several other places around Britain that make the same claim to fame. The story goes that Canute's courtiers were beginning to worship him as a godlike figure [he was named 'the Great']. As a deeply religious man, Canute knew this to be wrong so he went down to the river to show that he could not stop the rising tide. As he anticipated, 'his feet got wet' and this was symbolic of being powerless against the one true god [or nature]. But as we have learnt from the dangers of bores, surely if he was so close to the river – or even in the shallows – he would have been swept away by the turbulent currents?

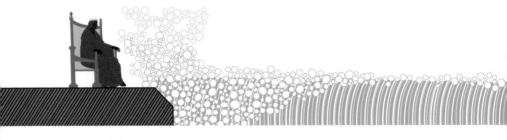

The bore travels to Britain's most inland port, Gainsborough

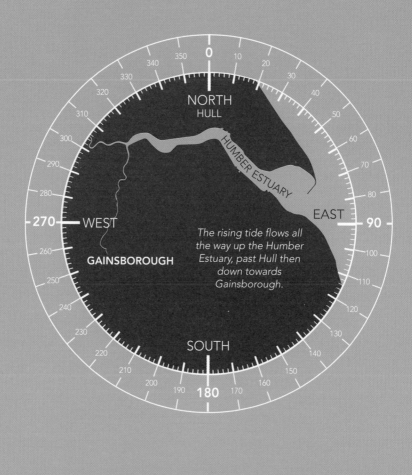

NORTH
HULL

HUMBER ESTUARY

WEST

EAST

270

90

GAINSBOROUGH

The rising tide flows all the way up the Humber Estuary, past Hull then down towards Gainsborough.

SOUTH

180

The bore travels 80 kilometres inland

The village of Arnside and its estuary sit in the north-eastern tip of Morecambe Bay. As we discovered in Chapter 1, the tide is famous here for racing in faster than a galloping horse and in Arnside this takes the form of a tidal bore. On spring tides over 9.5 metres, the leading edge of the rising tide is funnelled through the estuary into a 30cm-wave that rumbles past the village at 20 knots.

Thirty centimetres! Yes, it is small, but before you dismiss it against the leviathan Severn, think again. This bore is so dangerous, it has its own siren [established in 1969] to herald its arrival. Every summer, thousands of tourists with little or no knowledge of tides or bores flock to the sands at low tide. Because the bore is the first sign of the rising tide, the wave is often the only warning of its own danger. While the bore has the power to knock people off their feet, the real threat comes from being cut off by the fast rising tide that follows close behind [Morecambe Bay has a huge tidal range of 10 metres/30 feet].

The siren sounds twice. The first echoes across the estuary at around twenty minutes before the expected arrival of the bore. But as we have learnt in the past few pages – air pressure, wind and rainfall can all affect the time a bore arrives so this is an imprecise art. The second siren is more accurate because it is sounded when the coastguard spots the bore passing Blackstone Point. If you haven't done so already, this is the time to get to high ground or position your kayak to catch the wave [yes, it is a popular playspot].

The people of Morecambe Bay village want a siren, too. But the local council refused permission because it didn't want people to become too reliant on it and because the tide doesn't rise there as uniformly as a bore so it is difficult to pinpoint a particular moment when the bay becomes dangerous. So the locals are left to the method they have employed for years – blasting their car horns to warn unsuspecting tourists of the danger.

The bore arrives **around 2 hours before** high tide in Barrow

HIGH TIDE

-1   +1

NORTH

-2   +2

ARNSIDE

-3   +3

WEST   EAST

-4   +4

•BLACKSTONE POINT

-5   +5

SOUTH

-6   +6

The **30cm-wave** can reach speeds of **20 knots**

When we go down to the sea, the sounds are often very different from one day to the next. This could be down to the height of the tide; it may be windy or perhaps there are waves breaking. But when it comes to bores, the acoustic signature is constant and this is one of their key characteristics.

The defining sound element to a bore is its low frequency [human hearing range is generally 20 hertz to 16,000 hertz and bores can be as low as 50 hertz]. These low frequencies are synonymous with a deep rumble that can travel great distances without attenuation [reduction in level]. This explains why they are often compared to a booming bass or crashing thunder, and it also answers why bores can be heard long before the wave arrives.

The frequency of a bore has the potential to create vibrations within the chest cavity. With such resonance it is easy to see why the sound plays so deeply on human memories of the experience. Added to this, fear and awe have also been linked to low frequency sounds. Even animals struggle – they are often disorientated by the frequency of a bore and swept, fatally, into the waters.

The lowest frequency sounds of a bore come from entrapped air within the roller, as well as the turbulence caused by the breaking wave. There are also higher frequency sounds [around 150hz] and these are made by the water hitting the riverbed and banks. These can be divided into individual elements including the scouring of the riverbed, erosion of the riverbanks, the crashing of trees and water hitting jetties and man-made obstructions.

So when you next see a bore, think about these sounds and see if you can locate the different noises – it will be easier at night when there are less background distractions and no speedboats in the river. As human hearing usually deteriorates at high and mid-frequencies first, this can be a fun family outing. So you don't need to worry about grandpa forgetting his hearing aid, because this is one thing he will be able to hear.

Entrapped air in the wave   Outboard engines

Trees ripped out of banks   Scouring of the riverbed

With a plethora of tidal bores flowing up British rivers, it is easy to forget they are a rare phenomenon. There are less than 100 known bores in the world and I have only heard of one in the entire African continent [in Mozambique]. This shows just how lucky we are in Britain to enjoy these quirks of nature, especially in the north-west where two neighbouring rivers – the Dee and Mersey – each have their own bore that rumbles upstream at spring tides.

The River Dee is a beacon of cross-border unity. Its source is high up in the mountains of Snowdonia and the river meanders through Welsh countryside before crossing the border, passing through Chester and flowing into the Dee Estuary with Wales on one bank and England on the other. For its 110km length from source to sea, 16km is tidal and on spring tides a bore rumbles up this section.

The best bores develop along a 8km artificial channel built in the 18th century by engineers from the Netherlands. At the seaward end of this channel is Connah's Quay and the bore generally arrives here 2 hours before high tide Liverpool [which is around midday and midnight on 10-metre spring tides]. We have learnt in this chapter that wind, air pressure and rainfall can affect the wave so it is better to be half-an-hour early than two minutes late. If you do miss the wave by seconds, you can hop into a car and intercept the bore at Saltney Footbridge. It's a 10-minute drive, while the bore takes 30 minutes, so no need to rush.

Because the wave is less powerful than the infamous Severn, this is a perfect place to ease oneself gently into the sport of bore surfing. The average waves will be about 3–6m and they also travel slower than in the Severn. If you are surfing on a paddle-board or kayak, you can park at the Saltney footbridge and drift down with the natural flow of the river all the way to the Jubilee Bridge at Connah's Quay. When the bore arrives at 2 hours before high tide you can then surf it all the way back up to where you started!

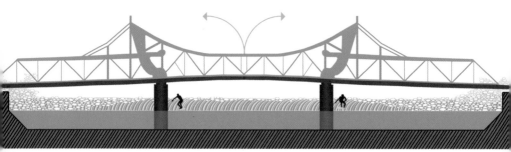

You can surf the bore for 30 minutes and travel 8 kilometres upstream

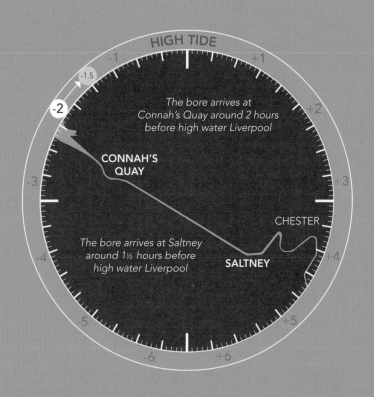

HIGH TIDE

-1
-1.5
-2
-3
-4
-5
-6

+1
+2
+3
+4
+5
+6

The bore arrives at
Connah's Quay around 2 hours
before high water Liverpool

CONNAH'S QUAY

The bore arrives at Saltney
around 1½ hours before
high water Liverpool

CHESTER

SALTNEY

The bore flows best up a straight manmade channel

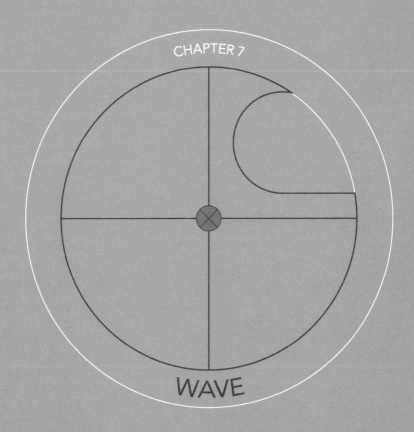

CHAPTER 7

WAVE

wave _a pulse of energy through water_

In this book, we have discovered moon-powered tidal waves, standing waves in rapids, tsunami waves generated by the sudden displacement of water and bore waves created by a rising tide. The wave we haven't explored is the one most people imagine when they think of the sea – the ocean wave. In this chapter, we'll simply call them waves, although their technical name is 'surface waves' because they are formed by wind blowing over the surface of the sea.

When a storm happens at sea, the friction between air and water molecules creates a wave that travels away from its source. Strong winds blowing over a large area for a long time make big and powerful waves capable of travelling across entire oceans. As they travel, the waves organise themselves into clearly defined groups known as sets, thus enabling even more power. These make great surf and people like me get very excited about a swell that has originated a long way from the local break.

The misconception with waves is that the water is flowing across the ocean. What actually happens is the water molecules flow in an orbital motion within one wavelength [the distance between two peaks] and the energy is transported through this motion. It is this energy that travels across the ocean, not the water. But when the energy reaches the shallows, the lower parts of the orbital columns of water are compressed and slow down. This forces the top to pitch up, and when the sea depth is 1.3 times the height of the wave, it breaks. If you're a surfer, that is the place you want to be.

I must warn you that this chapter may be a bit biased. When I was studying architecture at Newcastle University, we were set an essay challenging us to discuss a space that made us feel at home. I explained that surfing the glassy structure of a wave at nearby Tynemouth was an escape from the challenges of every-day life and left me feeling calm and invigorated – everything a home should be. Unsurprisingly, I was made to re-write the project and instead argued how my tent was 'home'. It comes as no surprise that five years later, I am a full-time camper travelling from one surf break to the next.

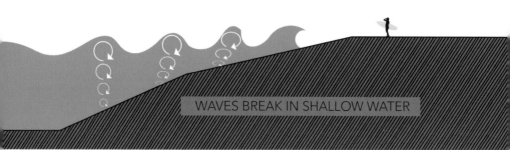
WAVES BREAK IN SHALLOW WATER

Kayakers are often found in the surf zone and the buoyancy of their vessel makes it easy to catch waves, but getting outback in big surf can be a challenge.

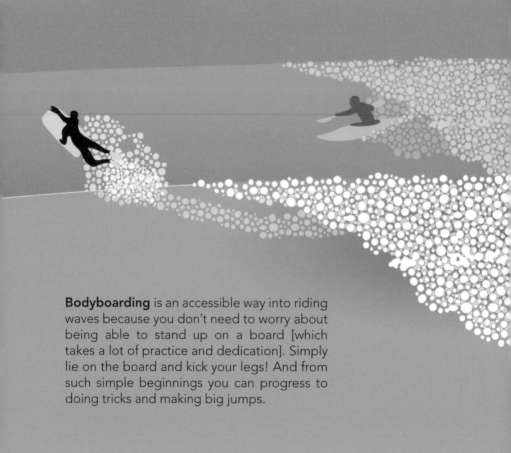

Bodyboarding is an accessible way into riding waves because you don't need to worry about being able to stand up on a board [which takes a lot of practice and dedication]. Simply lie on the board and kick your legs! And from such simple beginnings you can progress to doing tricks and making big jumps.

Stand up paddleboarding – or SUP – is similar to regular surfing but instead of paddling with your hands then popping up onto the board, you paddle while standing and you can use the paddle to help maintain speed on smaller and less powerful waves.

Surfboards are the most common way of riding waves and there's a board for every surfer and wave [see page 164].

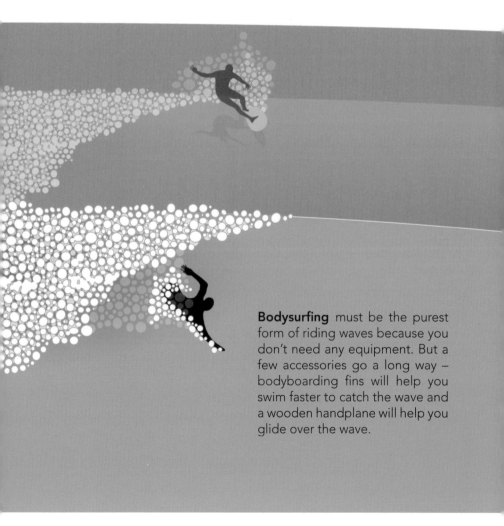

Bodysurfing must be the purest form of riding waves because you don't need any equipment. But a few accessories go a long way – bodyboarding fins will help you swim faster to catch the wave and a wooden handplane will help you glide over the wave.

Just like humans, waves come in a variety of shapes and sizes. If they have enough time before reaching land they will even organise themselves into groups. These are known as sets and for a wave to join a set it must have the same energy and travel at the same speed as the other waves. These two are directly linked to the period of the swell – the time in seconds between two consecutive waves. Generally, the longer and stronger the wind blows over the sea, the longer period that swell will have.

You can tell a lot about a swell by its period. Short period swells of below 10 seconds typically symbolise low energy and messy conditions. They cannot travel as far, have not had time to organise into clearly defined sets and are often found in the area where the winds are still blowing strongly. As you can imagine, none of these are good for surfers, so windswell is held in low regard. But as the period grows longer, more energy has been transferred from the wind so the waves will travel further, contain more energy and even grow taller when they reach shallow water.

Generally, doubling the period will increase the height of same-sized swell by 50%. As an example let's compare two 1.2 metre swells – one at 10-second period and the other at 20-second. The 10-second swell will grow to almost 2m waves on the beach but the 20-second swell will grow to 3m. And because long period swells [known as 'groundswells'] contain more energy, they can travel thousands of miles away from the storm that created them, thus increasing the chances of favourable offshore winds and better surfing conditions.

You can calculate the speed of waves by multiplying the period by 1.5. This shows that longer period waves travel faster and explains why big wave surfers need jetskis to catch the waves. Interestingly, in deep water, waves within a set all travel at twice the speed of the group. The back wave works its way forward, growing in size near the middle before diminishing until it disappears at the front. It then re-emerges at the back to replace the wave that has moved forward. Only when sets reach shallow water [half the depth of the distance between peaks] do they stop this phenomenon and travel at an equal speed to the group. This is when we like to surf them.

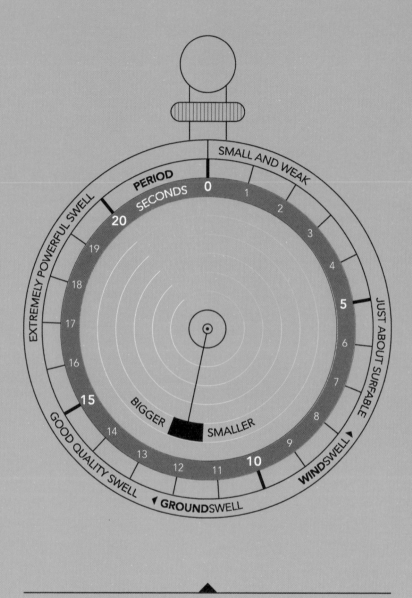

Waves in a set travel the same speed and share energy

When a wave approaches shallow water, the shape of the seabed has a profound impact on how it 'breaks'. For surfers this can make the difference between a gentle roller and a hollow barrel – as a rule of thumb, waves develop long and slow spilling forms over a shallow seabed. In contrast, if the seafloor rises steeply, the sudden change in depth makes the wave break over a short distance with a steep face.

This concept is simple in theory but in reality the underwater landscape is often irregular. This is where tide becomes important because the depth of water means the waves break above a different section of the seabed. On a flat beach, with steep banks nearer the shore, long and slow waves will form at low tide. But when the water level rises over the steep banks, fast and steep waves will fire out. Now I'm going to throw in a curve ball because the strong currents that accompany waves are always shifting the sand banks on a beach break. This means that from year to year the tide to wave combination can change – unless your break is a rocky reef that is impermeable to the ravages of the sea.

Reefs can be found both close to shore and far out to sea. They often signify a dramatic change in depth resulting in fast and steep waves. But if the tide is too high the wave may simply flow past without 'touching bottom'. If the tide is too low, the top of the reef may be out of the water, thus making it unsurfable. This means that many reefs can only be surfed at mid-tide. One that defies this rule is Porthleven in Cornwall, but its exception comes with serious dangers. At high tide there is a risk of being pinned within the cavernous structure, and at low tide there is so little water over the top of the reef that to fall off the wave would have the effect of jumping onto tarmac from a two-storey building. If you hit the rock you would then be at the mercy of 6 metre waves crashing down on you. This is why anyone below the level of professional is recommended to stay away at high and low tides.

Different shape waves can form at high and low tides

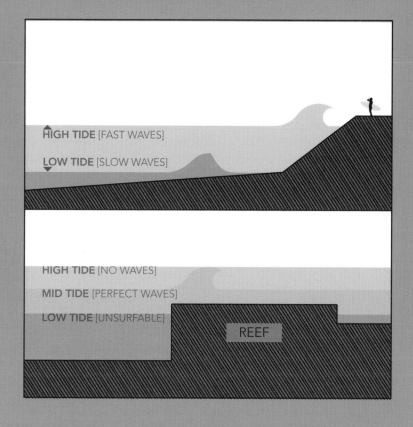

HIGH TIDE [FAST WAVES]

LOW TIDE [SLOW WAVES]

HIGH TIDE [NO WAVES]

MID TIDE [PERFECT WAVES]

LOW TIDE [UNSURFABLE]

REEF

A reef break may only form on mid or spring/neap tides

In the previous pages we learnt how a shallow beach with steep banks near the shore makes for gentle waves at low tide and heavy waves at high tide. This is the polar opposite to one of Britain's best beach breaks at Croyde Bay in North Devon. Here, steep banks further out to sea make fast hollow barrels at low tide. As the peak of the tide wave approaches from Lands End, the water above the banks becomes too deep for waves to form. Instead, the energy flows past and the waves peel into gentler rollers as the seabed shallows near the shore. This is a great time for beginners who may find the low tide waves too intense.

With dream conditions for all skill levels, it can come as no surprise that Croyde is a surfer's paradise. Yet it is a victim of its own success and the bay is infamous for overcrowding. This presents danger in the form of hundreds of pointy lumps of fibreglass flying around the place [every year a handful of people around the world die from head injuries sustained by out of control surfboards]. So how do you avoid the crowds? One option is to pop over the headland to the vast expanse at Saunton Sands. You will have plenty of space to yourself there and the flat beach means waves maintain their gently peeling faces for an eternity. This is perfect both for beginners practicing standing up and for experienced longboarders yearning to hang ten [gliding on the tip of their board with ten toes in the water].

What about the shortboarders who want a hard and fast barrel? My advice would be to come back in the winter when the swell is more consistent and the crowds have disappeared. But with the sea temperature dropping to a shivery 9° Celsius you're going to need a well-made 5mm-neoprene wetsuit with boots, gloves and hood. My favourite part of cold water surfing is watching dog walkers gasp in shock and awe as you nonchalantly paddle out into a snowstorm. The secret is to never let 'summer surfers' know quite how good the wetsuits are because; a] the waves will get crowded again, and; b] people will stop thinking you're tough by going surfing in a brutal British winter.

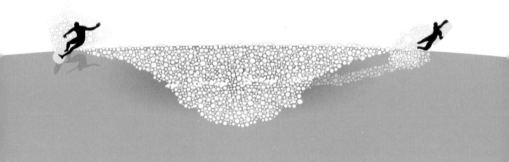

At **high tide** the waves are better for beginners

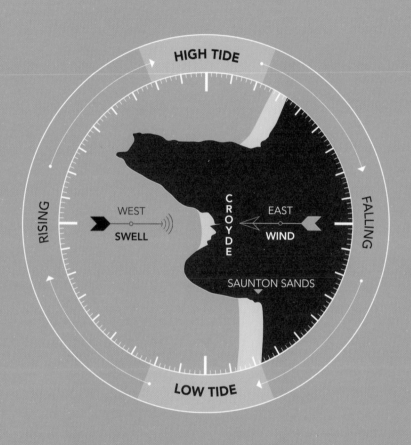

At **low tide** the waves are better for experienced surfers

I have quite a lot of boards in the camper. This is my 'quiver' and it consists of a shortboard, a mini-mal and an inflatable paddleboard [plus a longboard – the skate variety, and two more shortboards stored in my workshop]. Why so many boards? Well, as we just discovered in Croyde, the shape of waves can change from hour to hour – and that is only counting tide as a variable. When you add wind and swell [page 160] the conditions can transform from big to small, clean to messy, steep to shallow, fast to slow – the list goes on. To compensate for this, there is a board for every occasion...

Shortboard. This is what you see pros use when they make fast turns. The design has a steep rocker [the curve from top to toe]. This facilitates manoeuvrability but makes the board slow in a straight line. For this reason they perform best when ridden aggressively on powerful waves.

Longboard. As you would imagine from the name, this board is the polar opposite to the shortboard. The huge volume makes it easy to catch waves but the buoyancy makes it impossible to duck dive through big walls of white water when paddling out. These characteristics make them perfect for beginners on smaller waves. However, many experienced surfers only ride these boards because of the graceful form of surfing they encourage.

Mini-Mal. This is a blend of long and shortboards, thus providing a great transition between the two.

Fish. The name derives from the unique shape of its tail. A fish is often shorter than a short-board but the volume is compensated through width and thickness, making it very buoyant. Added with the flat rocker, a fish travels fast in straight lines [opposite to shortboard] and is perfect for typically small and messy summer surf because it glides over the white water to find the patches of unbroken wave.

Gun. This is recognisable by a distinctive narrow nose and tail that is designed to increase the length of rail in contact with the water, thus providing stability in really big waves. There is possibly only one place in Britain where you will ever need a gun, and that is the Cribbar [see page 176] near Newquay in Cornwall.

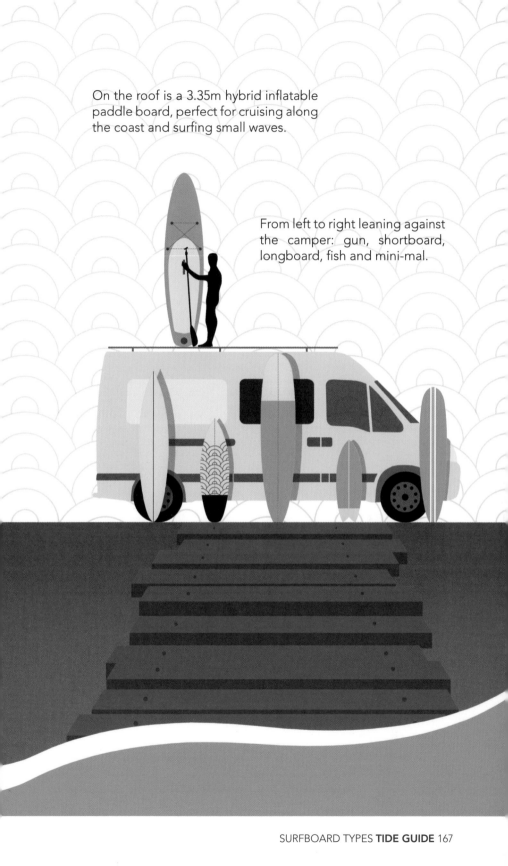

On the roof is a 3.35m hybrid inflatable paddle board, perfect for cruising along the coast and surfing small waves.

From left to right leaning against the camper: gun, shortboard, longboard, fish and mini-mal.

As we have discovered, the seas in the Pentland Firth are turbo-charged. This infamous stretch of water is home to some of the most powerful tidal streams in the world, as well as deadly rapids and fearsome whirlpools. Now we can add world-class waves to the list, because when powerful Atlantic swell pulses in from the north-west a volcanic reef at Thurso East moulds the energy into right-hand barrels worthy of surf magazine front covers. In case you were wondering – a barrel is a wave whose lip breaks over a surfer's head, and a right-hander breaks from left to right when you're surfing it.

The reef forms waves on all tides but it is best at mid-tide when the pumping barrel provides Europe's longest ride [excluding tidal bores of course]. The wave is so good that in 2006, Thurso hosted its first Cold Water Classic surf competition. For the inaugural event, the best surfers from around the world discarded their Hawaiian board-shorts for full body wetsuits and plunged into the icy cold waters. The look of shock was clear on their frozen faces. Who knew Britain had such good waves? Now, everyone does and Thurso has been dubbed our own 'North Shore' in response to Hawaii's famous surf breaks.

The Coldwater Classic put Thurso East on the world surf map, so now riders from Britain and overseas monitor the forecast with eager anticipation. When the swell starts pumping, the crowds head north. While the relations between locals and visitors are mostly respectful, reports of friction have trickled south with claims of visitors 'dropping in' on locals. This is the worst crime in surfing and involves paddling into a wave that someone has already claimed [the person surfing closest to where it is breaking has priority]. Scenes from the iconic surf film *Point Break* spring to mind, with an angry Californian cutting Keanu Reeves' leash with a penknife for committing such a cardinal sin. But such behaviour would never happen in Britain. That far into the wilds of Scotland, they're more likely to use a broadsword than a penknife – so don't drop in!

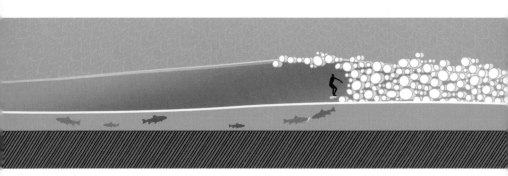

The waves can be triple overhead [6 metres]

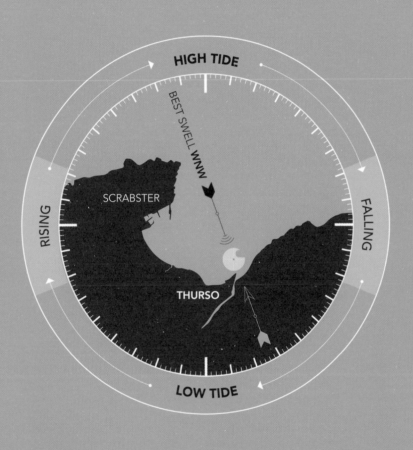

HIGH TIDE

BEST SWELL WNW

SCRABSTER

RISING

FALLING

THURSO

LOW TIDE

Enjoy Europe's longest ride at mid-tide

In August 2008, work began on Britain's first artificial reef, 225 metres from the beach at Boscombe, near Bournemouth. The plan was to lay down 55 geotextile bags on the seabed and pump them with sand. The biggest bag was 70m long x 2m tall x 6m wide and the shape of the reef was anticipated to double the height of the waves, their quantity, and the amount of people visiting the town to surf.

With the world watching in feverish anticipation, Britain's newest right-hand reef break was officially opened in November 2009 and the waves were … too difficult, too short and too infrequent. Disaster! All but the best surfers struggled with the hollow waves, and rides lasted far shorter than the 65 metres promised. The reef was a huge disappointment to everyone except the body-boarders who thrived in such intense, punchy waves.

The council had set eleven criteria to gauge the success of the reef and a report concluded that only four targets were achieved. As part of the contract, the company were called back to reduce the gradient of the ramp which would in turn produce shallower [easier to surf] waves. But soon after commencing improvement work, the company went into liquidation and the project was stalled. Now, the reef has been rebranded a 'coastal activity park' and the focus is on scuba diving and snorkelling.

This doesn't mean it is impossible to build a successful artificial reef. It just takes more planning and funding. To paraphrase one of the employees of the liquidated company 'it should never have been built there in the first place.' There is some good news though – the crabs, lobsters, pike, mullet and seabass are all very grateful for their new £3.2 million home.

The water is too deep at high tide to form waves

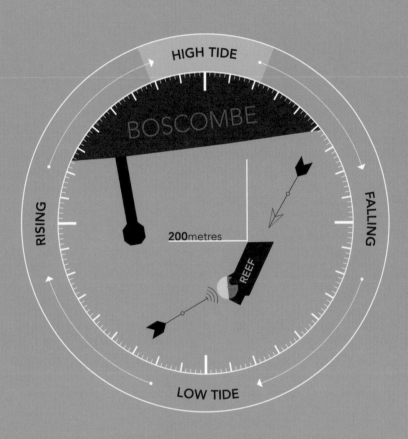

HIGH TIDE

BOSCOMBE

RISING

FALLING

200metres

REEF

LOW TIDE

The waves are too difficult, too steep and too infrequent

You can build the best artificial reef in the world but if there is no swell, there will be no waves. Engineers in Europe have overcome this problem by creating a lagoon with its own wave-generating machine, and the first of its type opened in the Welsh mountains in August 2015. The wave garden measures 300m long x 120m wide and has a central structure that runs down its length. Within this structure is an aquatic snowplough generating consistent head high barrels hour after hour. Heaven!

Separate zones within the lagoon offer waves to suit different skill levels ranging from 1.85 metres for experts to white-water rollers for beginners. The expert waves are so good that a well-known energy drink manufacturer hosted a competition where some of the best surfers from around the world battled against each other on the same number of identical waves. This must surely be the best way of testing board-riding skills without the inconsistencies of swell, wind and tide. However, being dependant upon technology, the wave pool has had its problems and closed twice in its first year due to mechanical errors. To fix the problem, 6 million gallons of water had to be drained from the lagoon. Nevertheless, this is to be expected when pioneering such exciting technology.

Could a perfect artificial wave ever replace the irregularities of surfing in the ocean? It is unlikely. I see it as the equivalent to a golf driving range. You go there to develop a specific element of your game within controlled conditions. Perhaps you are practicing popping up, cutting left or getting air – either way, you can repeat the manoeuvre over and over again on exactly the same shape, size and speed of wave. Then you can take your new skills and apply them to their true environment – the ocean. Out there, surfing is not just about riding waves. It's about tuning into nature. And there's nothing quite like the adrenaline and anticipation of not knowing when that big wave is coming or what's swimming around below. Overcoming these challenges and living to tell the tale is what being a surfer is all about and a wave garden can never recreate that.

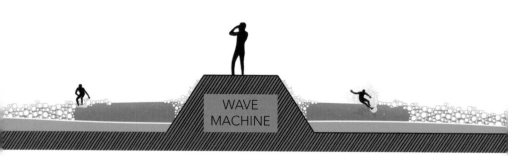

The wave garden produces **120 waves an hour**

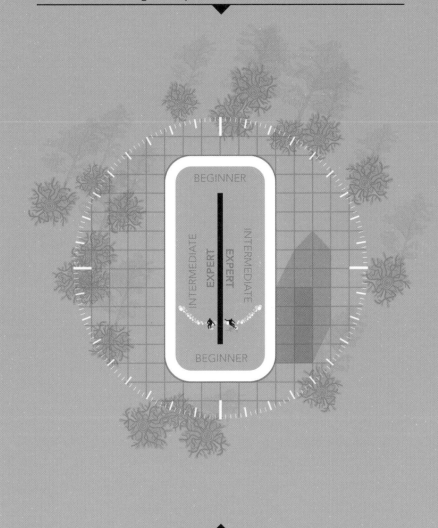

Surfers enjoy **18 second rides** on 1.85m tall waves

On the Cornish leg of our adventure around Britain, we very nearly got our camper permanently wedged within the narrow lanes of St Ives, so retreated to nearby Hayle in search of a beach to spread out. Following a network of roads in the general direction of the sea, we found ourselves at what looked like an electricity substation beside the beach. But this was no ordinary energy plant – this is wavehub, a testing ground for renewable energy. While the wave garden in Snowdonia consumes electricity to generate waves, wavehub generates electricity from waves.

An underground cable connects the power station to the actual wavehub – a 'socket' mounted on the seabed, 16 kilometres out into the Atlantic Ocean. This socket is positioned in the centre of a 4km x 2km rectangle, divided into quarters with each zone testing a different wave generator. The largest of the four schemes involves 200 generators mounted on the seabed and when it opens in 2017 it will become the world's largest wave energy park with a 10MW capacity. Each of the cylindrical generators is connected by tension tether [rope] to a buoy on the surface that moves up and down with the waves. This motion drives a piston within the generator to produce electricity.

Two of the four schemes are conceptually very similar to the first, but with a much larger flotation device secured just below the surface. The difference is that the kinetic energy is converted into electriciy within the framework. But all three designs face the challenge of how to protect the integral components from the ravages of the sea. This problem is overcome in the fourth design by locating the electricity generators back on shore. It also creates the electricity differently – by using the motion of the waves to pressurise seawater.

The high-pressure saltwater is then pumped back to the sub-station that doubles up as a desalination plant. By 2030, two thirds of the world will suffer from water shortages and taking the salt out of seawater is a solution to this problem. But the process is energy intensive, releasing thousands of tonnes of carbon dioxide into the atmosphere each year and worsening global warming. By using the energy from waves, this design not only helps solve the problem of water and electricity shortages – it does so without releasing any CO^2 into the atmosphere.

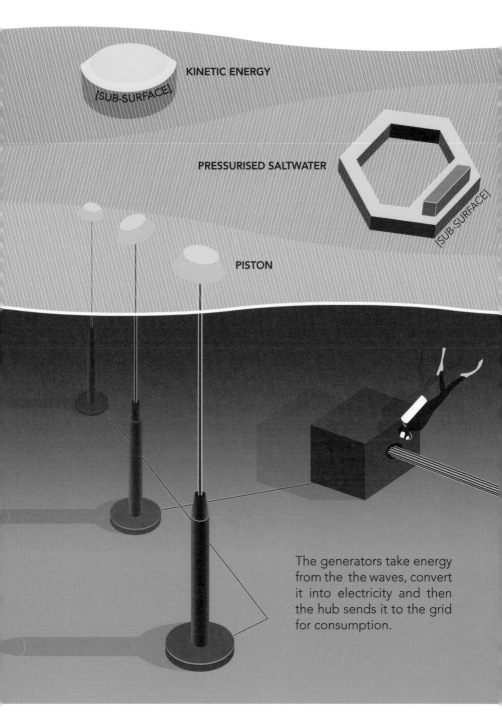

KINETIC ENERGY

[SUB-SURFACE]

PRESSURISED SALTWATER

[SUB-SURFACE]

PISTON

The generators take energy from the the waves, convert it into electricity and then the hub sends it to the grid for consumption.

When it comes to big wave surfing, most people think of Nazare in Portugal, Mavericks in California or any one of Hawaii's outer reefs. But right here in Britain, we have one or two spots of our own – the most famous being the Cribbar near Newquay in Cornwall. This is a reef just off Towan Head and it only breaks around once a year when the swell is devastatingly powerful. But for the conditions to be surfable, the wind [both speed and direction] plus tide must be perfectly aligned with the swell. And even then, the reef break is so dangerous it is nicknamed 'the widow maker' – powerful rips [see next chapter], long hold-downs and razor sharp rocks are just some of the dangers.

According to local folklore, the Cribbar was first surfed by three Australians in 1966. But it was not until the turn of the century that British surfers frequented the reef on the rare occasion it was surfable. Nowadays, most surfers use a jet ski to tow them into the wave. Not only does this guarantee they can generate enough speed to catch the wave [long-period swells travel fast] – it's a handy rescue device when they find themselves in the danger zone between house-sized walls of white water and the bone-breaking rocks of Towan Head.

The underwater profile of the seabed can be simplified into three tiers. Because bigger waves break in deeper water, the 'smaller' 4.5m swell breaks on the shallowest tier just off the headland. Anything smaller will simply crash onto the rocks. But as the swell gets bigger it breaks further out, with the biggest waves breaking 400 metres out to sea with 12-metre faces. You might think someone must be slightly unhinged to put themselves in such a dangerous environment – maybe, but they must also be exceptionally skilled. Either way, the Cribbar seems to be the limit of British big wave surfing and nobody has taken the challenge to the next level – the Zorba reef. This rocky shelf lies 3km out from the Cribbar and makes even bigger faces than the widow maker. And, as yet, it has not been surfed. Could you be the first?

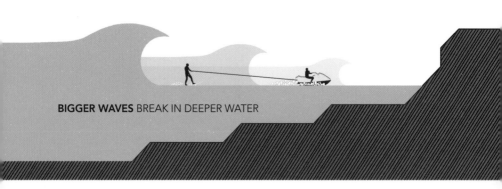

BIGGER WAVES BREAK IN DEEPER WATER

Surfers need jet skis to tow them into 40 foot waves

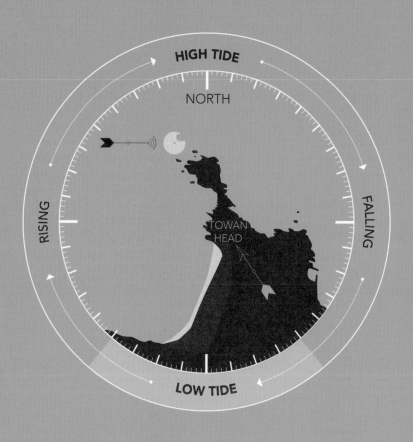

HIGH TIDE

NORTH

RISING

TOWAN HEAD

FALLING

LOW TIDE

The big wave spot is only surfable on average once a year

A big wave like the Cribbar will break with a force of around 6 metric tonnes per square metre. That's a tremendous amount of energy, but nothing compared to a rogue wave. These monsters of deep water are classified as being twice the height of the average top third of waves in a swell and they release a boat-breaking force of 100 tonne/sqm. For modern vessels built to withstand an intense 15 tonne/sqm, the power of one of these freak waves is simply too much. Furthermore their unpredictability catches seafarers by surprise – the waves can break out in the open ocean with very little warning. The good news is that scientists are currently developing computer systems to predict a rogue wave approaching, allowing ships to alter course or offshore rigs to batten down the hatches and clear the decks.

For centuries, sea captains told stories of waves with 30-metre vertical faces, but people struggled to believe them. Ernest Shackleton spoke of a gigantic wave on his epic Voyage of James Caird from Elephant Island to South Georgia in 1916. In 1980, the biggest British ship to be lost at sea, the 91,655 tonne *MV Derbyshire*, was believed to have been sunk by a rogue wave 300km off Japan. And, in 1995, the *QE2* reportedly surfed down the face of a 25-metre wave in the North Atlantic. Despite these reports, it was not until the first official recording in 1995 that scientists started to invest time and money in researching rogue waves.

This first accepted reading was measured from the Draupner Platform in the North Sea. That wave was recorded at 26m and damage to the structure at that height supported the data. This proved to the world that these killer waves really do happen, leading scientists to develop two main theories explaining what they call 'extreme storm waves'. The first suggests that because waves travel at different speeds, there is a short time when two or more converge and their shared energy creates an extra large wave. The second theory argues that because rogue waves often form in areas of strong ocean currents [such as the Gulf Stream] the interaction of wave against current steepens the wave face. This would explain the common description of rogue waves having a deadly combination of extremely steep faces and terrifyingly deep troughs.

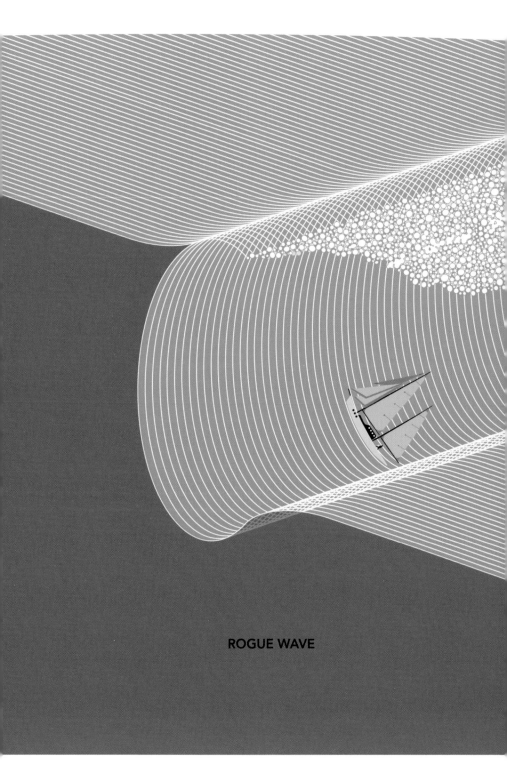

ROGUE WAVE

Heading back to the coast and putting rogue waves behind us, what makes perfect surfing conditions? One main factor is wind – the movement of air from areas of high to low pressure. When hot air rises, it creates low pressure at ground level and wind blows in from colder areas to replace the lost air. This process can happen on a global stage, but here we will explore the phenomenon on a much smaller scale – the sea breeze.

As the sun warms us throughout the day, land and sea heat at different rates. Because the sea has a greater ability to absorb the sun's rays, it heats up and cools down slower than land. Starting at dawn, the land is colder than the sea so wind blows away from the coast to replace the rising sea air. This is an offshore wind and helps create 'clean' surfing conditions. The wind blowing against an incoming wave helps hold up the face, allowing it to travel into shallower water and develop a steeper form. This explains why surfers head to the beach at the crack of dawn.

The early morning offshore breeze doesn't last long because when the sun rises it starts to re-heat the land. By mid-afternoon the land is far hotter than the sea and this creates low pressure over land and high pressure over sea. The colder sea air then blows towards the coast to plug the gap. This is an onshore wind and often brings messy surf conditions with choppy 'windswell' waves mixed in with the long-distance groundswell. Because the wind is blowing with the waves, they break sooner without developing the peeling forms common with offshore winds.

There are exceptions. If an offshore wind is blowing greater than 40kph it deteriorates the conditions. It also makes it difficult to gain speed because you are surfing into the wind. While most surfers avoid onshore winds, professionals make the most of such unique conditions to practice aerials. Not only does the breaking of the wave help them get air, the wind direction helps their feet stick to the board when airborne. And because there are fewer surfers in the water, there is the added bonus of not worrying about landing on anyone.

Wind blows from areas of high to low pressure

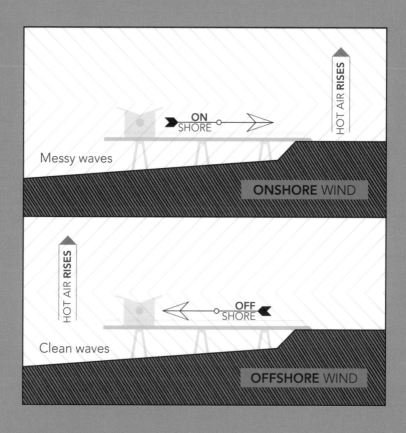

Think of wind as the aerial version of eddies in tidal rapids

Tynemouth is the home of surfing in the north-east and the place I learnt to surf [when I should have been in architecture lectures at nearby Newcastle University]. And what a baptism of fire that turned out to be! My toughest session came one afternoon in late October when I spent the better part of an hour paddling into the cold fury of the North Sea without getting anywhere. Stumbling back onto the beach with exhausted arms, this would have been an appropriate time to quit the ridiculous sport. But I was determined to ride waves so returned when my arms had recovered and the swell was pumping [which is only 31% of the time in peak winter months]. This time I tried a different tactic and waited until a break between sets, then paddled out as hard as I could. After just a few minutes I was beyond the breakers, enjoying my first lesson in seamanship – learning to take when the sea gives and stay well away when it takes.

One year on, my surfing had improved and winter was approaching once more. As the short days closed in, my friends and I found ourselves surfing long after sunset, until one day we took it to the next level and decided to go night surfing. Looking back, this was ridiculously dangerous and could easily have escalated into a life-threatening situation. But at the time it was a natural step in our thrill seeking, so we paddled out into the darkness. In truth, it was never fully pitch-black because the bright lights of Tynemouth dazzled over the surf. But when we looked out to sea there was nothing but darkness. When a wave approached, the only indicator of its arrival was a streak of white at its lip. One, two, three paddles and we would drop down the face, waiting to reach bottom and discover how big the wave was.

The scariest part of night surfing in Tynemouth was getting tumbled around underwater with no sense of which way was up. During the day sunlight filters from the surface but in the pitch-black of night we would have to climb up our leashes in the hope they were still connected to our boards on the surface. Then there were the rips. We'll learn about them in the next chapter.

Tynemouth is suitable for **all skill levels**

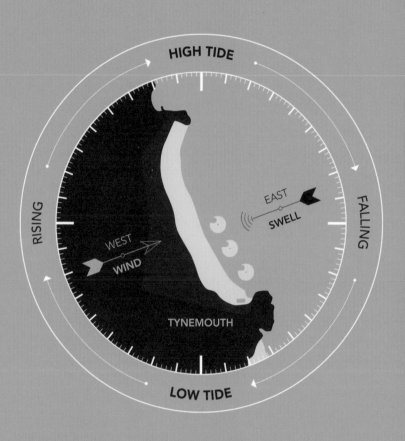

HIGH TIDE

RISING

FALLING

EAST

SWELL

WEST

WIND

TYNEMOUTH

LOW TIDE

The beach break can produce good waves on **all tides**

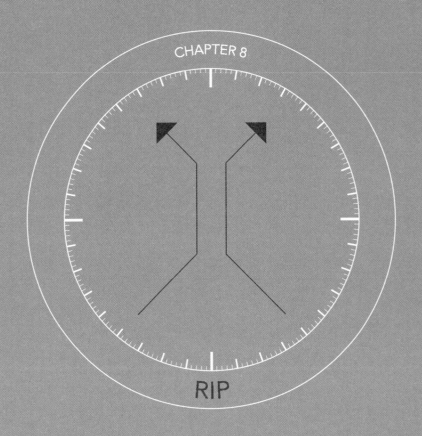

CHAPTER 8

RIP

rip *a flow of water out to sea*

If there is just one chapter you read in this book, this should be the one – because around 2,000 people are rescued and at least 20 drown in rip-related incidents on British beaches every year. This makes them the deadliest motion of water against our island, for the simple reason that many people lack one basic piece of knowledge – never swim against the current. A recent study showed that only 25% of beachgoers knew how to spot a rip and how to escape if caught in one. This is a shockingly low statistic and I can guarantee that by the end of this chapter you will understand not only how to escape a rip, but how to have fun with one, too.

A rip current is a narrow channel of water [usually less than 10 metres wide] that can flow hundreds of metres out to sea, beyond the surf zone. The surf aspect is important to understand because rips and waves go hand in hand. As water surges up the beach with a breaking wave, it flows along the shore until it finds the path of least resistance back out to sea. This can be a gap between underwater sandbanks, a deeper channel within a reef, or obstructions such as sea defences or headlands. When the water reaches one of these bathymetric or topographical features, it flows straight out to sea at a rate of up to 2.5 metres a second. That will take you a hundred metres from the beach in less than a minute.

For surfers – and anyone else wanting to get 'out back' beyond breaking waves – rips are a fantastic way of doing so quickly and effortlessly. But for someone caught unsuspecting, being dragged out to sea comes as a terrible shock and can induce panic. Without basic understanding of these currents, people don't know what to do so they try to swim straight back to shore. But not even an Olympic swimmer can make headway against this flow, and people lose their lives by swimming without getting anywhere, becoming exhausted and then drowning. To make sure this never happens to you, keep reading.

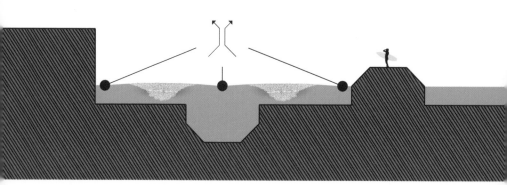

In Perranporth, we moored our 'land yacht' [commonly known as a camper van] on a hill above the beach and settled down for the afternoon work session. With the sliding door wide open, I watched in fascination as rips ebbed and flowed with the tide [it was not a very productive work session]. The type of current I was watching was a sandbar rip – the most common in Britain. It forms when waves break on the beach, causing water to flow along the shore until it reaches a gap between the sandbars where it flows out to sea. I noticed that as the tide was falling, the rips were building in power – this is because the shallower water over the banks increases wave action and forces water through the gaps at greater speeds.

Newspaper headlines every summer report 'MASS RIP RESCUE' from Perranporth, usually with around 15 people plucked from the sea by RNLI lifeguards. These mass rescues often happen on multiple beaches along the Cornish coast at exactly the same time, usually when low tide coincides with a powerful set of long-period waves hitting the shore. When these high-energy bodies of water break on the beach, the sudden outward flow of water creates 'rip pulses' and the speed of flow can jump from around 0.5 mps to 2.5 mps. Because the wave surge momentarily increases the depth of water, people are knocked off their feet and become more susceptible to the current. NOTE: if you are in a rip and can stand up, do so, because it makes you more stable against the flow of water.

The sensation for most people caught in a sandbar rip is being pulled straight out to sea. But recent research using GPS floats at Perranporth has shown that this phenomenon only happens around 20% of the time – it is known as a 'rip exit' – and water flows straight out to sea then dissipates beyond the surf zone. For the other 80% of the time, the rip moves in a circulatory pattern, just like an eddy. If you are in this channel and simply float, you will find the water will take you out to sea, then parallel to the shore, then back in towards the beach with the waves. This revelation has changed the way we approach escaping rips, but it has also added a new complication because different tactics are better depending on whether you are in the 'rip eddy' or the 'rip exit' – and it is often difficult to tell.

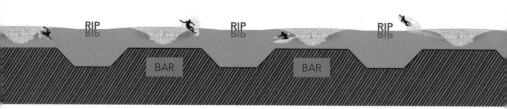

Gaps between sandbars are the most common cause of rip

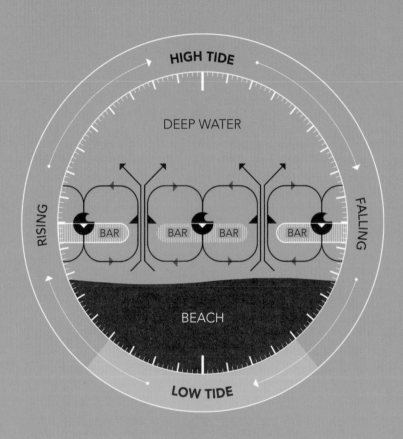

The rips are generally more powerful at low tide

The best way to spot a rip is to find some high ground above the beach and wait for a set of waves to arrive. As we have discovered in the last few pages, rips are linked to waves and bigger waves make powerful rips. But you may have to wait a while, so in the meantime analyse the features of the beach where rips may form. If it is low tide, look out for gaps between sandbars and bear in mind that rips will happen as the tide rises over them. If there is a river mouth, the outflow of water will have carved a deeper channel and rips will form there. In the next section, we will also learn about how obstructions such as headlands, groynes, sea walls or piers also create these deadly currents, so be aware of these.

Once you have ascertained where to look for rips, it helps to know how to spot one. The most noticeable feature is a gap between waves – this signifies the deeper water where rips flow. As the water is moving out to sea, it will be discoloured with sand or sediment dragged out from the beach. The most obvious sign is someone caught in one [call 999 and throw in something floaty – going in to help them is a last resort because the rescuers often end up being the ones who drown. If you do go in, take something buoyant and be prepared for irrational behaviour because people panicking at sea often use their rescuers as a flotation device and force them underwater].

Understanding the individual features of a rip will help you spot one quicker – the three components are the feeder, neck and head. The feeder is the section nearest the beach and this is often shallower than the water around it. Water from the feeder is fed into the neck and this is typically the narrowest and fastest element of the rip. It is often darker than the water around it because of its greater depth [if it is a sandbar rip]. As the neck passes beyond the surf zone it dissipates into the head with a distinctive arc shape of foam or sediment gently merging into the clearer water around it. Spotting just one component of a rip will help you piece together the other elements, but remember there are usually multiple rips on a surf beach, so keep looking. And because they ebb and flow with the tide, they can change on an hourly basis, so keep a regular lookout.

Ironically, people often head straight to the rips to bathe because they think the calmer water is safer than the waves.

When a wave breaks at an angle to the beach [45 degrees is optimum], it generates what is called a longshore current and this flows along the coast until it is deflected out to sea by a coastal structure such as a headland or groyne. This channel of water is known as a topographical rip and kills more than ten people around Britain every year. Although sandbar rips are more common, topographical rips are more powerful from the same sized waves and they can travel out to sea twice as far beyond the surf zone. This presents a lethal danger for someone caught in one without the basic understanding of what's going on.

Topographical rips can form when waves are as small as 0.5 metres and with 75% of British beaches adapted with coastal structures [33% have groynes, 55% have headlands, 60% have natural geological outcrops] there are plenty of places for them to form. Because they happen on low-energy surf beaches, they are common in the English Channel and the North Sea where people are less familiar with rip currents so don't know how to escape them. These coastlines are also susceptible to strong tidal streams and although it has not been substantiated by research, I wonder if this flow could replace or add to the longshore current in generating these rips. The tide itself can also affect the rip because if the groyne is out of the water at low tide, no rip will form. But as the sea level rises and more water comes into contact with the structure, the rip will grow.

The shape and positioning of the obstructions plays an integral part in how topographical rips form. Studies carried out at Boscombe beach have shown that if the spacing between the groynes is less than four times their length, the rips will rapidly diminish in size. If the spacing is two times the length of the groyne, the rip speed will be 25% less than if the spacing is four times. This is because there is not enough space for the longshore current to generate speed and feed the rip. When the water hits the groyne, the further it flows along the structure, the exponentially longer the rip will form. Groynes that extend beyond the surf zone can generate rips twice this distance – far longer than sandbar rips that dissipate just beyond the breakers. So, unless you want to find yourself in deep water - watch out for any structure heading out to sea.

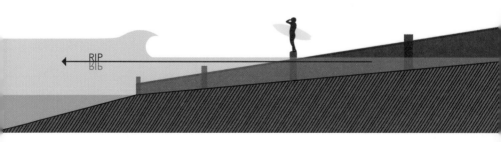

Topographical rips are more powerful than sandbar rips

▼

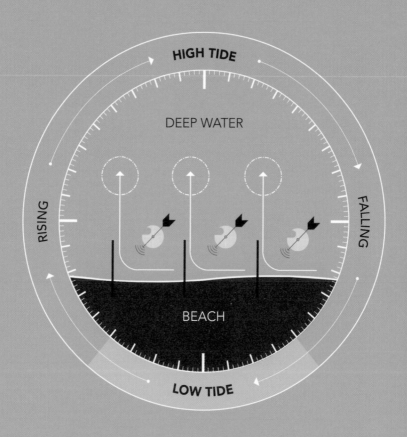

HIGH TIDE

DEEP WATER

RISING

FALLING

BEACH

LOW TIDE

▲

When the groyne is fully exposed at low tide, there will be no rip

We explored the Gower Peninsula near the end of our circumnavigation of Britain and it was mid-summer. Walking down into Three Cliffs Bay – the most photographed location in the Gower – it was difficult to resist the temptation of diving into the cool waters. But we had passed a sign warning of dangerous rip currents, so severe that it was recommended not to bathe at all. In the past six weeks, two men had drowned on this very beach, in almost identical ways. It was mid-afternoon and their children were playing in the shallows when they were knocked off their feet and swept into the rip. The fathers dived in to rescue them and lost their own lives. Both men were fit and strong, but as I've said even the world's greatest swimmer would eventually drown in a rip if he tried swimming against it.

There are two main reasons why so many people drown in Three Cliffs Bay. Firstly, few people fully understand how the rips work here. In embayed beaches with two headlands, the circulation patterns are more complex. Also a river weaving through the bay provides a deeper channel for rips to form, adding to the complications. I visited a similar beach on my European travels and a large sign indicated where the rips form – I have applied this theory to the map but do take it with a pinch of salt. What really needs to happen is a thorough research study on the motions of water within this deadly bay. In the meantime, the RNLI has recently set up a station on the beach providing cover on weekends and holidays between the spring and autumn.

The lifeguards constantly assess the conditions, putting out red and yellow flags to signify a safe swimming zone free from rips. Red flags indicate it is too dangerous to swim. This service will undoubtedly save lives, but the station has only been established since the men drowned. So when their children got caught in the rip they had very little options and as a parent, I would have done the same thing. But long before we get into such a traumatic situation, I would ensure that my children are fully aware of what rips are, how they work and how to spot and escape them.

BREAKING WAVES ARE THE
SAFEST PLACE TO SWIM

The rips at Three Cliffs Bay mostly form at high tide

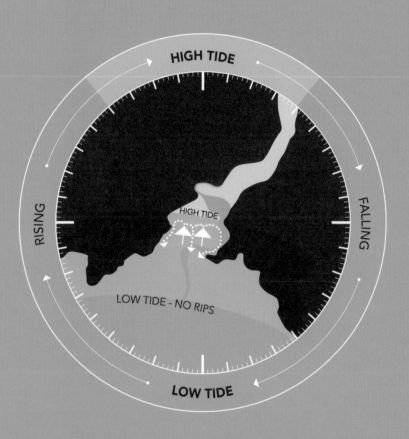

HIGH TIDE

RISING

FALLING

HIGH TIDE

LOW TIDE - NO RIPS

LOW TIDE

Embayed beaches make complex rip current circulations

The number one rule to surviving a rip is **never swim against the current**.

Escaping a rip starts on the beach and the first thing to do is find some high ground and look out for the signs of currents. Locate the features where you know rips will form and look out for their component parts – feeder, neck and head. But these are sometimes difficult to spot and they can quickly turn on and off with slight changes in tide and even wave angle.

If the beach is patrolled by the RNLI, it is sensible to swim between the red and yellow flags. That way if you get caught in a rip all you need to do is float and wave your arms. The lifeguard will be looking out for exactly this scenario and by staying calm you will be making his or her job of rescuing you very simple. However, lifeguards only patrol a small proportion of our coast and they close down in October for the winter.

So if you find yourself caught in a rip on a glorious late autumn day, you may have to rescue yourself. If you are in the middle of a long sandy beach you can be confident that you are in a sandbar rip and there is an 80% chance the current will recirculate you towards the waves and back into the beach. But you need nerves of steel to simply lie back and go with the flow – especially if this is your first time in a rip and you know there is a 20% probability you are in the 'exit channel' that will deposit you straight out beyond the surf zone. So I would advise you to gently swim across the current towards the waves – especially if you are in a topographical or mega-rip.

The people who drown in rips are generally those least familiar with these currents. So when I am surfing I actively find and use rips as much as possible – they are a great way of getting out beyond the breaking waves. Also by maintaining my skills within these currents, I am more likely to stay calm in the chance I do get caught in one by surprise when swimming. Remember that staying calm is the most important aspect to escaping these currents because the opposite mindset – panic – is what kills people.

Do not bathe in the calm areas between waves – these are rips

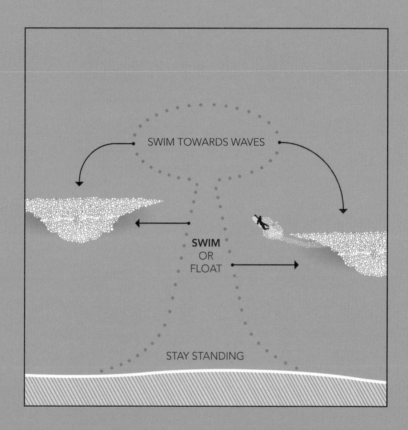

SWIM TOWARDS WAVES

SWIM
OR
FLOAT

STAY STANDING

In the feeder, try to stay on your feet and wade out

The main danger of rips does not necessarily come from the power of water [unlike tsunamis and whirlpools] but from people's lack of understanding. Being caught in a rip does not mean you are going to die: being caught in a rip without knowing what to do is what leads to drowning. For every one hundred people on a British beach, seventy-five do not know how to spot and escape a rip. As I mentioned earlier, this is a shocking statistic, but it may come as a surprise that not even scientists studying rips can yet claim to fully understand the inner working of this motion of water. Research has come a long way in the past decade, but there is still a lot to learn.

Scientists study rips by simulating ocean conditions in a laboratory wave tank, through mathematical and computer modelling and by carrying out live experiments in the surf zone. In Perranporth, this started by surveying the seabed and working out where the sandbars are and how they change with time. This involved zigzagging along the beach at low tide on a GPS survey enabled quad bike. Then the equipment was mounted onto a jetski with sonar and this developed a detailed picture of the seabed. Once the information was plugged into a computer, the researchers mounted instruments on the seabed measuring wave height, period and angle. Then they released floats with GPS trackers and studied the circulatory patterns of rips and how they ebb and flow with changes in the tide, seabed and waves.

In Poole Bay, a yellow dye was released into the water upstream of a groyne to visually demonstrate the movement of water particles within a topographical rip. Across the Atlantic in North America, scientists have even dredged a 45m x 27m x 1.8m channel in a beach to better study how rips work. Because the principles are consistent throughout the world, we can apply this research to British beaches. And I can imagine it won't be long before we can watch the forecasted hourly flows of rips on our smartphones. The intention is not to change or control rips – they are an essential element to the beach ecosystem – but instead to live more harmoniously with these motions of the sea.

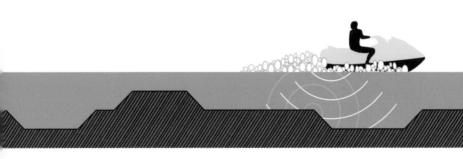

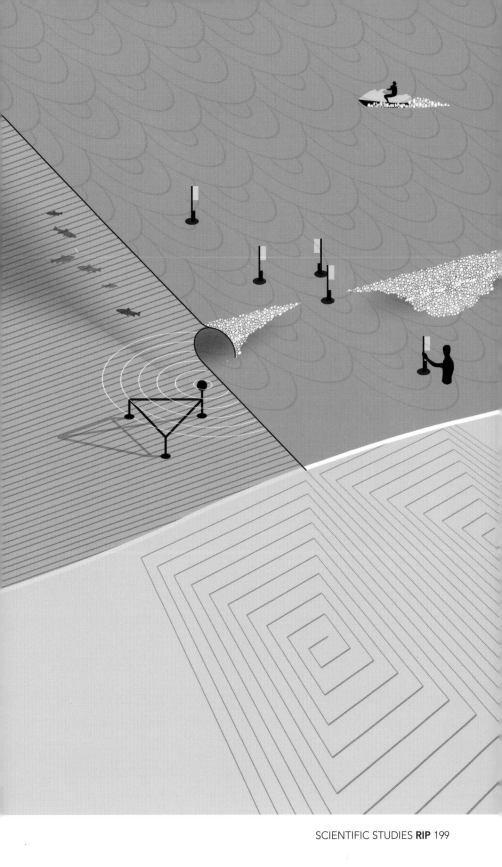

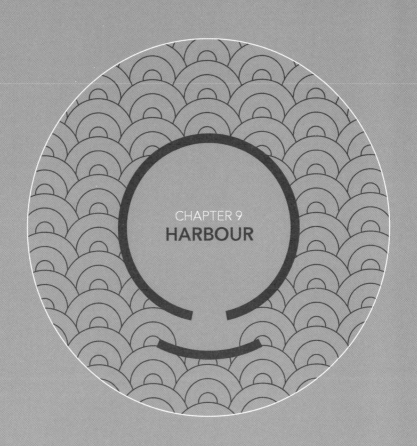

CHAPTER 9
HARBOUR

Our journey through the coastal waters of Britain has left me with an unusual feeling of being invigorated and exhausted at the same time. I imagine you feel the same, and hope that [like me] you have learnt something along the way. So let us find a space free from the motions of the sea, somewhere to rest and absorb everything we have just discovered. We need a harbour.

Harbours come in all shapes and sizes but their basic function is simple. They should provide a consistently calm environment free from tide, stream, rapids, whirlpools, tsunami, bores, waves and rips. But not all harbours achieve this and my experience has shown that many dry out at low tide, waves batter others and some even have their own tidal stream cycles. In the extreme events of tsunami, colossal whirlpools have been known to form in big harbours and many ships have sunk in rivers because bores took them by surprise.

Harbours are more than breakwaters, dredged channels and tidal gates – they are the heart and soul of coastal communities. They provide a safe space for fishermen to unload their catches, and for sailors to take refuge from storms. Strategically located harbours have helped us win wars, but they have also enabled invading powers to maintain a grip on power by providing a safe supply base. Successful harbours have brought economic prosperity through trade with other nations. Even today a majority of our goods are shipped by sea – I believe developing harbours that can provide consistent conditions for ships to efficiently unload cargo will keep Britain economically competitive in the future.

Once we have re-energised from our time in harbour, let us go on one more adventure. But instead of searching for the motions of water as we did before, this time let us look for the places where they do not exist – if there is such a utopia.

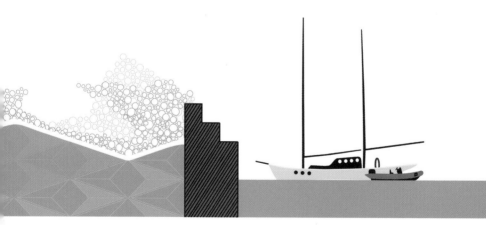

The Downs is an anchorage off the coast of Deal, where our camper adventure around Britain began. An anchorage is the loosest form of harbour and is simply a body of water sufficiently suited for ships to drop anchor. During the days of sail there would have been hundreds of large vessels – both Royal Navy and merchant ships – anchored here. But without jetties and harbour walls, everything had to be transported from ship to shore by small boats that could be hauled up onto the pebble beach. As a result, the town of Deal developed into England's busiest port in the late thirteenth century, as it handled the movement of goods and people to and from the ships.

The success of the Downs is directly linked to the Goodwin Sands. Lying to the east, the sands act like a breakwater and absorb the energy of waves out at sea. Land to the north and west also help enclose the anchorage. With an area of 129 square kilometres and depths up to 20 metres, the Downs is capable of sheltering a huge number of large ocean going ships. The highest reported number of vessels at one time was eight hundred. All these ships were here for one main reason – to wait for favourable winds to continue their journeys north towards the Thames or south into the Channel and beyond.

The greatest weakness of the Downs is a gaping hole in its defences to the south. Adverse wind and waves from this direction penetrate straight into the anchorage, and this was highlighted to devastating effect in the Great Storm of 1703. An extra-tropical cyclone came in from the south-west and combined with powerful tidal streams that flowed through the Downs, tore many ships from their anchors and swept them onto the Goodwin Sands. One thousand men died on the sands that night.

As ships propulsion systems were adapted from sail to motor, there was no longer a need to wait for favourable winds. So they became redundant, except for the occasional ferry or cargo ship awaiting entry into Dover Harbour on stormy days. Nowadays the most sail action you can see from the beach is a handful of dinghies from the appropriately named Downs Sailing Club, battling both powerful tidal streams and each other to win their weekly race around the buoys.

The anchorage can be used at all tides

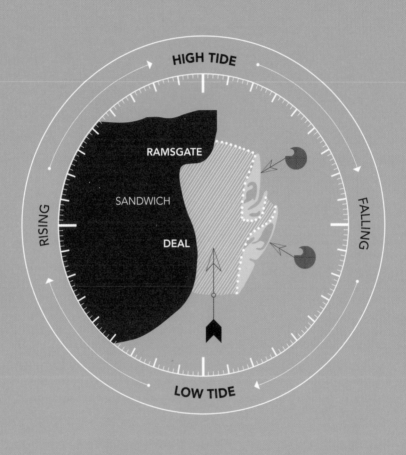

HIGH TIDE

RISING

FALLING

RAMSGATE

SANDWICH

DEAL

LOW TIDE

Its weakness is a lack of defence in the south

Just a few miles south-west of the Downs lies Dover Harbour, Europe's busiest ferry port. When roll-on/roll-off ferries were introduced in the 1950s they expected ten thousand cars and motorcycles in the first year but it was so popular one hundred thousand arrived. Now five million vehicles pass through every year.

The harbour as we see it today, with its three outer walls – the Admiralty Pier, Southern Breakwater and Eastern Arm – was completed in 1909. There are two entrances – the eastern and western – and the western is a particularly dangerous place to be on a small boat in rough water at high tide. As swell crashes against the breakwater the surge is deflected back into the incoming seas and this creates steep and deep waves coming from all angles. The powerful tidal streams flowing along the coast make it difficult for low powered vessels to enter the western, even on calm days. I was once providing safety boat cover for a cross channel rowing boat when they were swept across the entrance and could not actually get into the harbour.

When tidal stream is flowing south it pours directly through the eastern entrance, creating a unique and complex stream system within the harbour. But this poses little challenge for large ships – especially when local tugs guide them in – and a £120 million Western Docks Revival Scheme is developing a new cargo distribution centre to attract more cargo ships using the harbour.

The regeneration of Dover Harbour comes at a price, because to undertake this enormous construction project they need to dredge the Goodwins for sand. Not only are the Goodwin Sands of historical significance – they provide a habitat for a colony of seals – but as we learnt in the last page, they provide a breakwater for the Downs and that makes them a natural sea defence for local coastal towns. Dredging the sands may be a cost-efficient form of rebuilding Dover Harbour but many locals – myself included – fear this will come at great cost to the marine ecosystem and the communities who already suffer from considerable beach erosion.

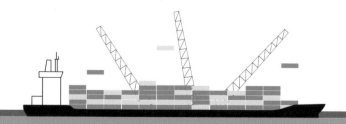

The harbour has its own complex tidal stream system

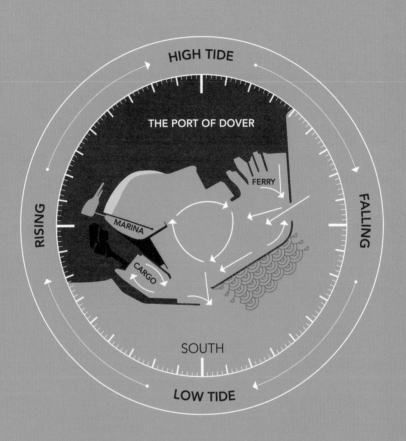

The harbour covers an area of 283 hectares with depths of 10m

Poole Harbour is the biggest natural harbour in Britain [and one of the largest in the world] and at its peak was the principal port trading with North America. This enriched the economy and 90% of the local labour force was dependant on the harbour. But in 1847 a new railway line coincided with the development of deeper draught ships [the average depth of the harbour is just 48cm] so trade quickly shifted to the deep-water port at Southampton. Poole suffered as a result. However, an Act of Parliament in 1895 created the Poole Harbour Commissioners and they set about rejuvenating the area.

By the time we moored our camper at Sandbanks – in May 2015 – the harbour was booming once more. The clear blue waters shimmered on the sandy seabed and I excitedly inflated the paddleboard to join an armada of kite-surfers, wind-surfers, wake-boarders and yachts. This was too much adventure for Alfie [our water spaniel] to sit by and watch so he dived in beside me and we paddled around the lagoon together. But we had to keep our wits about us, because Royal Marines often practice parachuting into these sheltered waters. There are also big ferries and freight ships to avoid, attracted by a constantly dredged 7.5-metre deep channel from the main quays to the harbour entrance. For all this activity it may come as a surprise that Poole Harbour is also a place of worldwide importance for wildlife conservation. Feeding, roosting and breeding birds all thrive on the mudflats and salt marsh, with endangered red squirrels living on the islands.

Poole sounds like the perfect harbour, given it provides shelter for both humans and wildlife. Its geography even reduces the tidal range to almost non-existent levels. But this phenomenon is linked to an effect known as a double high tide, which complicates the stream cycle at the entrance. Usually water flows for six hours in either direction but in Poole it flows in for four hours, out for four, in for one and a half then out for three. When water from the 36 square kilometre harbour drains through the single entrance it can reach speeds of 5 knots. This has resulted in many small boats being pinned against the iconic chain ferry and in extreme cases they have even been dragged under. Luckily the crews have all managed to re-surface downstream – albeit a little shaken up.

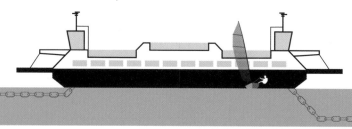

This is one of the biggest natural harbours in the world

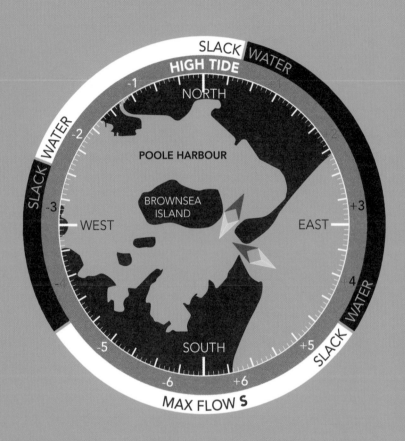

Powerful and complex streams flow through the entrance

The road to North Berwick took us onto a high plateau bordering England and Scotland. The sun was setting and wind turbines spun all around us, flickering in the golden light. The road itself was dead straight, tar black and seemed to be leading us directly into the burning sunset. It was an epic moment. Then the plateau suddenly dropped precipitously to the sea and we zigzagged down the steep slope, stopping to watch a fleet of fishing boats returning from the North Sea far below. In the distance was the silhouette of Law Hill and North Berwick, our destination.

We arrived later than expected because of a slight detour via the less glamorous Berwick-Upon-Tweed. My sister was to blame, because she said we simply must visit 'Berwick', forgetting the North bit [it actually just used to be called Berwick until they added North to avoid such confusion]. So it was dark when we arrived, eager to discover what lay outside. As usual, Alfie woke me up in the morning and we jumped out of bed onto the beach where there was a tidal pool, filled each day by the rising tide and offering swimmers the chance to bathe in pure seawater without worrying about stream, waves, rips and rapids – a harbour of sorts.

Our view out to sea was dominated by the dramatic Bass Rock, rising 100 metres from the Firth of Forth just offshore. It appeared to be snow-capped, or made of white rock, but with disbelief we realised it was covered in birds. Northern Gannets to be precise – this is the biggest colony [150,000] of them in the world, and you can take a boat trip out to see them. The trips only leave the twelfth century harbour a few times a day and it's clear why – all the boats are lying in the mud.

Even though the harbour was deepened in 1804 and again in 1831, it dries out at low tide. This is a common weakness in British harbours and must be so frustrating for businesses when their boats are stuck in the mud, especially when all other conditions beyond the harbour are perfect. But it is no trouble for us – I work in 3-hour periods so by the time I finish making a new tide map the harbour is filling up again. We climb aboard a catamaran and race out to Bass Rock with puffins flying along beside us.

The harbour was built in 1150 and deepened in 1804 and 1831

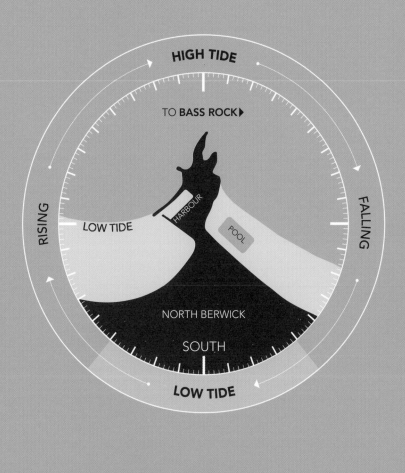

HIGH TIDE

TO **BASS ROCK** ▶

RISING

FALLING

HARBOUR

LOW TIDE

POOL

NORTH BERWICK

SOUTH

LOW TIDE

The harbour completely dries out at low tide

Let us go north, further than we have been before – to Scapa Flow, an almost 380-square kilometre anchorage sheltered from the Atlantic and North Sea by the islands of mainland Orkney, Graemsay, Burray, Hoy and South Ronaldsay. Although this one billion cubic metre body of water may seem overly remote, history has shown it as a harbour of national strategic interest. The Vikings first utilised Scapa Flow as an invasion base one thousand years ago and when World War One broke out the British Grand Fleet moved in. At the end of the war the German High Seas Fleet was brought here but they scuttled 54 warships instead of allowing them into the hands of the British [although all but 7 were recovered in the greatest salvage operation of all time].

The naval tactic in World War One was to control the North Sea and cut off German supply lines. Scapa Flow was the perfect place to do this from, while sheltering from the ferocious streams, rapids, whirlpools, waves and rips of the Pentland Firth. A drawback to the harbour was its plethora of entrances between the islands and the Admiralty was anxious about submarine attacks, so they sunk ships in the entrances and tied nets between them. Tide and stream could freely pass through but German submarines could not.

When World War Two broke out, the Royal Navy returned to their safe haven at Scapa Flow. But in the intervening years the submarine defences had fallen into disrepair and on 14 October 1939, *U-47* slipped through Kirk Strait and sunk the colossal HMS *Royal Oak*. Over 800 sailors were lost. This devastating attack led to Winston Churchill personally ordering four of the eastern entrances to be entirely filled in – these double up as roads and you can drive over the 'Churchill Barriers' today.

As World War Two came to an end, the Royal Navy left Scapa Flow. Now its strategic importance comes from the oil and gas pipeline from the North Sea, which supplies half of Britain's energy demand. But this industry is in rough waters so Scapa Flow may lay silent once more, waiting patiently for when we need it next.

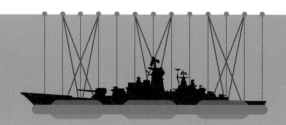

Scapa Flow has space to harbour an entire Navy

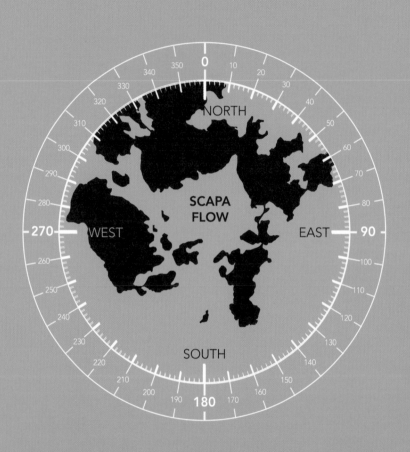

SCAPA FLOW

NORTH

WEST

EAST

SOUTH

There are one-billion cubic metres of water in Scapa Flow

COLD WATER

FACTS

SEA TEMPERATURE

The cold waters around Britain can induce hypothermia.
Survival time without a wetsuit is one hour in 10 degree water*

*however, you can acclimatise your body to cold water immersion

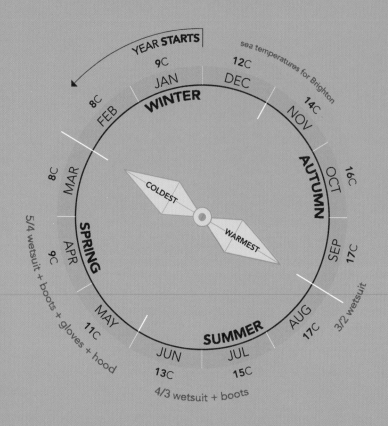

YEAR **STARTS**

sea temperatures for Brighton

9C
JAN
WINTER

12C
DEC

8C
FEB

14C
NOV

8C
MAR

16C
OCT

SPRING

AUTUMN

9C
APR

17C
SEP

5/4 wetsuit + boots + gloves + hood

11C
MAY

17C
AUG

3/2 wetsuit

SUMMER

13C
JUN

15C
JUL

COLDEST

WARMEST

4/3 wetsuit + boots

WETSUIT SELECTION

5C - 11C

11C - 15C

15C - 20C

5/4

4/3

3/2

* a 5/4 wetsuit means the torso neoprene is 5mm-thick and extremeties are 4mm

DEPTH GAUGE

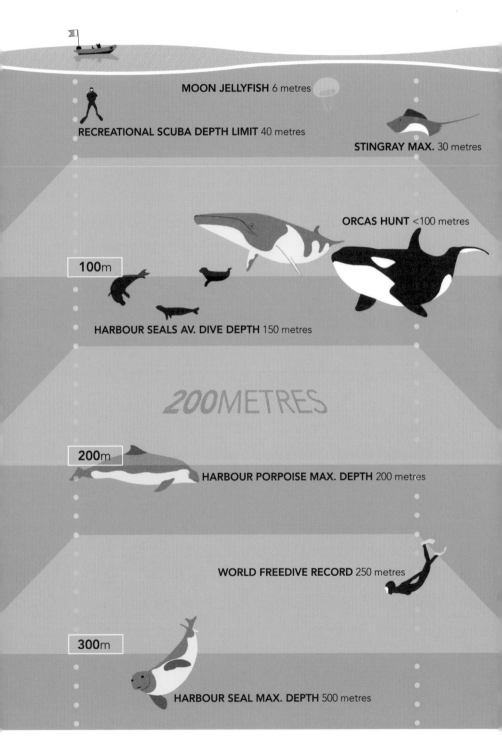

MOON JELLYFISH 6 metres

RECREATIONAL SCUBA DEPTH LIMIT 40 metres

STINGRAY MAX. 30 metres

ORCAS HUNT <100 metres

100m

HARBOUR SEALS AV. DIVE DEPTH 150 metres

*200*METRES

200m

HARBOUR PORPOISE MAX. DEPTH 200 metres

WORLD FREEDIVE RECORD 250 metres

300m

HARBOUR SEAL MAX. DEPTH 500 metres

CO$_2$

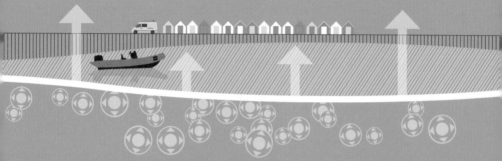

Carbon emissions around the world are damaging our coast.
We can help by reducing demand on burning fossil fuels*

*oil, gas and coal

RISING SEA LEVELS

Carbon dioxide emmissions are warming the earth's surface and the sea around Britain is heating up. As the water particles heat they expand, and if you combine this with melting ice the result is an increase in the high tide level of our coastal waters – possibly by up to 2 metres by the end of the century, although more dramatic forecasts suggest a 7-metre rise.

This increase in the sea will result in more damaging storm surges as there will be a higher platform for the sea to rise from. This will increase flooding of low-lying coastal areas, damaging agricultural land and contaminating aquifers with saltwater. And it is not just humans who will suffer – fish, birds and plants will lose their habitats.

How can you reduce this? It's simple – convert your energy demand from fossil fuels to renewables.

SCIENTIFIC RESEARCH

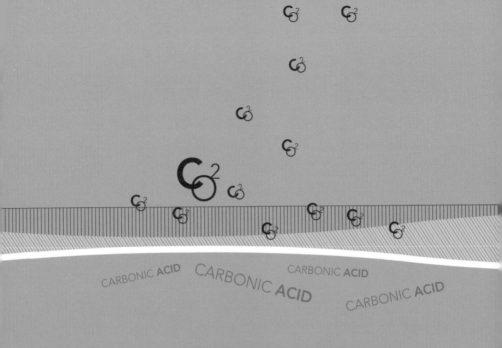

WATER ACIDIFICATION

The sea absorbs half the carbon dioxide from our atmosphere and when this reacts with seawater it creates carbonic acid that is making Britain's waters more acidic. In fact, our sea is 25% more acidic than 200 years ago [the beginning of the Industrial Revolution when we accelerated the burning fossil fuels to power machines], after millions of years of pH stability.

The increase in acidity is restricting the growth of shells – coral, lobsters, oysters, plankton - and can damage fish reproductive systems. In short, it's killing sea life around Britain.

Britain is home to a deepwater coral-making reef – **Lophelia** – supporting a rich ecosystem of sealife, but acidification is threatening this habitat.

PLASTIC

Manmade materials such as plastic are finding their way into the coastal waters around Britain and it's damaging wildlife. From seabirds ingesting plastic to orcas washed up on the beach after being tangled in rope, managing waste responsibly is vital to maintaining a healthy marine environment around our coast.

PLASTIC BAGS look like jellyfish and a tasty meal for many marine life, including seals. They last 20 years in the sea and release toxins while the plastic breaks down.

PLASTIC BOTTLES take 450 years to break down in the sea and while this is happening they are a danger to fish who swallow the small pieces of plastic.

FISHING LINE is found in the stomachs of 90% of fulmars [a type of seabird] that are found dead on the beach. The line takes 600 years to break down.

WILDLIFE SPOTTING

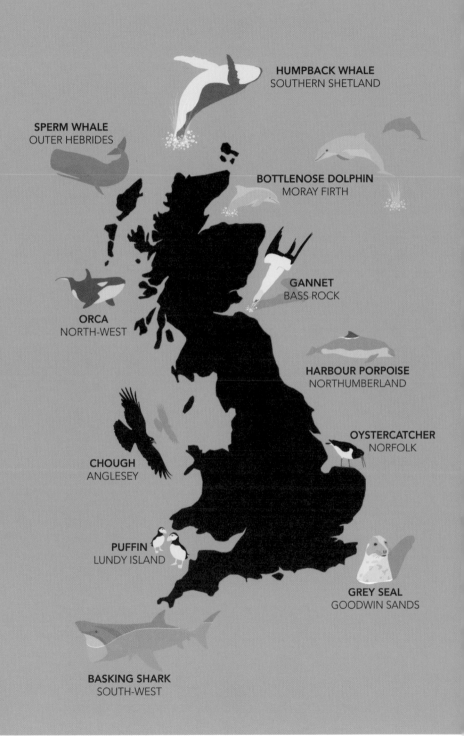

HUMPBACK WHALE
SOUTHERN SHETLAND

SPERM WHALE
OUTER HEBRIDES

BOTTLENOSE DOLPHIN
MORAY FIRTH

GANNET
BASS ROCK

ORCA
NORTH-WEST

HARBOUR PORPOISE
NORTHUMBERLAND

OYSTERCATCHER
NORFOLK

CHOUGH
ANGLESEY

PUFFIN
LUNDY ISLAND

GREY SEAL
GOODWIN SANDS

BASKING SHARK
SOUTH-WEST

TIDE

TSUNAMI

SUN

WHIRLPOOL

CAMPER

BUOY

WAVE

1 NAUTICAL MILE
1.15 LAND MILES
1.85 KILOMETERS

WETSUIT

MINKE WHALE

TIDAL TURBINE

SUP'ER

RIP

DINGHY

SHAG

BORE

1 KNOT=1 NAUTICAL MPH

RAPIDS

POWER BOAT

LAND YACHT